MICROWAVE VALVES

C. H. DIX, B.SC., M.I.E.E.

W. H. ALDOUS, B.SC., A.R.C.S., D.I.C., M.I.E.E.

Research Laboratories of the M–O Valve Company, Ltd.
G.E.C. Hurst Research Centre

MICROWAVE VALVES

LONDON ILIFFE BOOKS LTD

First published in 1966 by
Iliffe Books Ltd, Dorset House
Stamford Street, London, S.E.1

Printed and bound in England by
W. & J. Mackay & Co. Ltd., Chatham, Kent

CONTENTS

PREFACE

Valves are an interesting group of active electronic devices which generally have the function of transforming electrical energy, mostly from d.c. to pulsed or alternating energy, but occasionally in the other direction also. In many of these functions, particularly at low-power levels, valves have been supplanted in the past ten years by solid state devices (and no doubt this trend will continue, so that some of the devices discussed herein may become obsolete). However, for the generation and amplification of power at microwave frequencies (say 1,000 Mc/s upwards) valves are at present the obvious choice and usually the only possibility.

In this book we are principally concerned with microwave valves, which are distinguished from their lower frequency counterparts by the fact that the r.f. circuits are actually part of the valve, and that transit times through them occupy many r.f. periods. Many books have already been written about microwave valves, but most of these are far more advanced and specialised in their scope than we have attempted to be here. What we have tried to do is to present an account which can readily be absorbed by the technically educated reader e.g. of graduate or H.N.C. level. Some limited mathematical treatment has been included in those places where it is felt that it adds to understanding, rather than being necessary in order to enable quantitative estimates to be made. The approach nevertheless aims at being a fundamental one, starting with the existence of the electron and the behaviour of electro-magnetic fields, and covering briefly what is necessary to develop the ideas used in microwave valves, passing through many of the aspects of conventional or space-charge control valves on the way.

Despite their inconvenience when dealing with electrons, the conventional directions of current flow and electric field are used, i.e. electric fields are in the opposite direction to the forces they produce on electrons, and current flow is in the opposite direction to electron motion. In most places, and whenever no other units are specified, the MKS system has been used.

During the last ten years, much work has been published describing new microwave devices or systems of interaction, methods of electron beam focusing etc., but where this work has not resulted in marketed devices, we have not included it. No doubt some of these ideas will in time result in new available devices, but we do not feel equal to the task of looking into the future and deciding which will succeed, and generally the book is limited to discussing devices at present available.

Wembley, C.H.D.
January, 1966. W.H.A.

Chapter 1

INTRODUCTION

1.1 Electricity, Matter and Electrons

Although it may safely be assumed that any readers of a book such as this will be already familiar with most of the principles and results of Electricity and Magnetism, it is perhaps worth while to recapitulate, in what may be a slightly unusual sequence, some of the essential aspects of electromagnetism which are involved in valves.

About electricity it is difficult to say anything fundamental. It is an attribute of matter, or a form of energy; it is the glue that sticks molecules together and the stuff that makes telecommunications possible.

Matter, the solid, liquid and gaseous stuff of which the earth is made, is composed of molecules, each of which we think of as being the smallest particle which still has the properties of the whole. A molecule of water is still water, anything less is not. The molecules in turn are composed of atoms of the elements of which the molecule is made up; again in the familiar case of water, two hydrogen atoms and one oxygen atom making up the molecule of water.

Atoms, in their turn, are made up of still smaller particles, which for the present we shall regard as the fundamental particles. These are protons, neutrons and electrons. The overwhelming majority of the mass of the atoms is in the protons and neutrons, while the electron, which is the particle in which we have the greatest present interest is more noteworthy because of its electric charge.

A normal hydrogen atom, which is the simplest of all atoms, consists of one proton and one electron. The proton has a mass of $1{\cdot}67 \times 10^{-27}$ kilogramme and a diameter of about 10^{-14} metre, and carries a positive electric charge of $1{\cdot}6 \times 10^{-19}$ coulombs. The electron, whose mass is only about 1/1840 of that of the proton, carries an equal negative charge, and orbits around the proton at a distance of about 5×10^{-11} metre. Overall, therefore, the atom is uncharged, since the charges of the proton and electron neutralize each other outside the atom. The next more complicated atom, of helium, has a nucleus of two neutrons which are uncharged, and two protons, and therefore weighs about four times as much as a hydrogen atom and has twice its charge. It then has two electrons in orbit around the nucleus, to produce an overall neutral atom. Atoms of other elements follow in similar succession until we arrive at the largest and most complicated of all naturally occurring atoms, that of uranium, one of whose possible atoms has a nucleus containing 92 protons and 146 neutrons, giving a total mass of 238 times that of a hydrogen atom. It has 92 orbiting electrons to give it an overall zero charge.

1.2 Electron Currents in Solids

It will, of course, be known that electric current is the passage of electric charge. The only electric charges that exist are those of protons and electrons, if we neglect the rarely occurring positron, or positive electron, and sub-atomic particles. This charge is therefore the ultimate fundamental unit of electricity. It is not possible to have any smaller charges. An electric current therefore must consist of the movement of electrons in one direction or of protons in the other direction. Sometimes, as in electrolysis, these will carry with them the ions, or charged molecules to which they are attached or of which they form part. In general, in thermionic valves we are only concerned with electrons, so for the time being let us forget about the other much bigger and heavier particles.

So far, all the electrons that have been encountered have been bound up with nuclei in atoms or molecules. It is possible to detach electrons from an atom, and in a large class of solids, which because of this property are called electrical conductors, electrons are more or less free to wander through the solid. If, for instance, we take a long thin electrical conductor, i.e. a wire, and join its ends to the terminals of a battery, which for the moment we can regard as an electron-pump, there is a continuous flow of electrons round the closed circuit through the wire and through the battery, which we recognize as a current flowing in the wire. If, by using a large battery we make the electron current big enough, its passage will cause the wire to become hot, as in an electric fire, due to the buffeting of the atoms of the wire by the moving electrons. The distance that any electron can run without colliding with an atom is very small; in a copper wire, for example, it is only about 10^{-8} metre. The electrons can be thought of as behaving like small ball bearings running through an assembly of much larger balls, which are relatively immovable. These impede the flow of the electrons along the wire, and the battery has to supply the force to drive them through, thereby doing some work. This work appears as heat in the wire, the quantity W in joules being given by the well-known relation

$$W = VIt$$

where I = current in amperes
V = battery voltage
t = time in seconds.

It is worth while obtaining some idea of the magnitudes involved in current flow through a piece of copper wire. Since the negative charge on an electron is $1 \cdot 6 \times 10^{-19}$ C, then for a current of 1 ampere (=1 C/s), $6 \cdot 25 \times 10^{18}$ electrons must pass any point in the wire each second. Now copper has $8 \cdot 35 \times 10^{28}$ atoms/m³, so that in a 1 m length of 20 gauge (0·914 mm diam.) wire there will be $5 \cdot 5 \times 10^{22}$ atoms. With one free electron per atom, this means that the electrons in this size of wire only have to travel at $1 \cdot 14 \times 10^{-4}$ m/s to give a current of 1 A, or to put it another way, it would take approximately 2½ h for the electrons to travel 1 m.

1.3 Electrons in Vacuum

Before leaving this section, we shall just mention the phenomenon of *thermionic emission*, which will be returned to later. This is that if a wire is heated, either by the passage of an electric current or in any other way, the increased molecular agitation causes some electrons to be thrown right out of the surface of the metal. If the wire is enclosed in a vacuum, the electrons can then move quite large distances without colliding with anything, and so the study of their behaviour becomes much easier. It is with electrons, moving in vacuum-filled enclosures, that we are mostly concerned in dealing with valves.

1.4 Electric Field and Potential

The electric charge of an electron is discovered (or defined) by observing its effect on other electrons; that is to say that in the space around every electron in vacuum there is an *electric field* whose main characteristic is that it exerts a repulsive force on any other negative charges, i.e. other electrons near by. Like the gravitational field, this electric field obeys an inverse square law; it decreases rapidly as the distance from the electron increases. If we make the electron move against this field, it has to do work, and hence lose kinetic energy just as a smooth ball loses energy in rolling up a smooth hill; if the hill is high enough, the ball will be brought to rest, and then roll backwards. An electron meeting a retarding region of electric field behaves in the same way, being brought to rest and then accelerated backwards. In the case of the ball, the force on it at any point is proportional to the gradient of the surface at that point, that is to say the maximum rate of change of the height of the surface with distance. This gradient is a vector quantity, it has a magnitude or size, and a direction, which is the direction of the steepest slope at the point. It is therefore more complicated and difficult to use in calculations than the height from which it is derived, and so in general, wherever possible, we work with the height rather than the gradient.

In electricity, the quantity corresponding to height is potential, or voltage, and the electric field is the gradient of the potential. Writing this mathematically, we say that if the potential increases by a small amount δV in a small distance δr, the electric field is given by

$$E = -\frac{\delta V}{\delta r}$$

The minus sign expresses the fact that the direction of a field is defined as the direction in which a positive charge would move, which is away from the high potential, not towards it.

1.5 Acceleration of Electrons by an Electric Field

The force that an electric field exerts on an electron is proportional to the field, and to the charge on the electron. It is thus given by

Force (in newtons) = Electric field (in volts per metre)
× Electronic charge (in coulombs)

or
$$F = -Ee$$

This time the minus sign comes from the fact that the charge of an electron is negative. Taking this with the previous equation, it can be seen that wherever there is a variation in voltage from one point to another an electron will be accelerated from the low-voltage region to the high-voltage region.

Work equals force × distance, so if an electron moves the small distance δr in which the voltage increases by δV, the work done on it by the electric field is

$$\left[e \times \left(\frac{\delta V}{\delta r}\right)\right] \times \delta r$$

that is $e\delta V$. This gives an increase in the kinetic energy of the electron; hence if an electron starting from rest is accelerated through a voltage V, its kinetic energy ($\frac{1}{2}mv^2$) is given by

$$\tfrac{1}{2}mv^2 = eV$$

where m is the mass of the electron ($9 \cdot 11 \times 10^{-31}$ kg). This produces a new unit for the measurement of kinetic energy, the *electron volt*. An electron with an energy of 1 electron volt is moving at a speed of $5 \cdot 94 \times 10^5$ m/s.

1.6 Stored Energy of an Electric Field

We have seen that an electron produces around it an electric field. This is, in fact, the only way in which an electric field can be produced. If it is desired to produce a very large electric field, or a very high voltage, this can only be done by assembling a large number of electrons together and this must involve some expense of energy, since each electron added to the assembly has to overcome the force due to the field which already surrounds it. It follows therefore that the existence of an electric field always represents some stored energy. A simple example of this is that of a charged capacitor. A large excess of electrons, having a total charge Q coulombs is induced on the surface of the capacitor plates by connecting them to a voltage source of V volts. This produces a strong electric field between the plates, and the energy stored in this field, which is the work done in charging the capacitor, is

$$\tfrac{1}{2}QV \text{ joules}$$

1.7 Magnetic Fields

The electric field due to an electric charge is there whether the charge is moving or not, but in addition, if it is moving, it also produces a magnetic field, and, moreover, just as all electric fields are associated with local collections of electrons, all magnetic fields, even those of permanent magnets, are associated with movements of electrons. There is, however, a difference in that whereas the electric field radiates from the electron, and all static electric fields must start from and end on charges, magnetic fields form closed rings around the direction of the movement. The strength of the magnetic field is, as will be known, proportional to the current, i.e. the rate at which electrons are passing. This is the product of the electron density and electron velocity. In metallic conductors, the electron density can be enormously high, and so even although the velocity is low, large magnetic fields can be readily be produced. In vacuum, however, where there are no positive protons to neutralize the repulsion of the electrons for each other, it is very difficult to achieve high electron densities, and so large magnetic effects only occur if the electrons are moving very fast.

It can now be seen that this picture of electric and magnetic fields is just the same as that learnt at school. If electrons are moving along a conductor of length *d*, connected as shown in Fig. 1.1, across

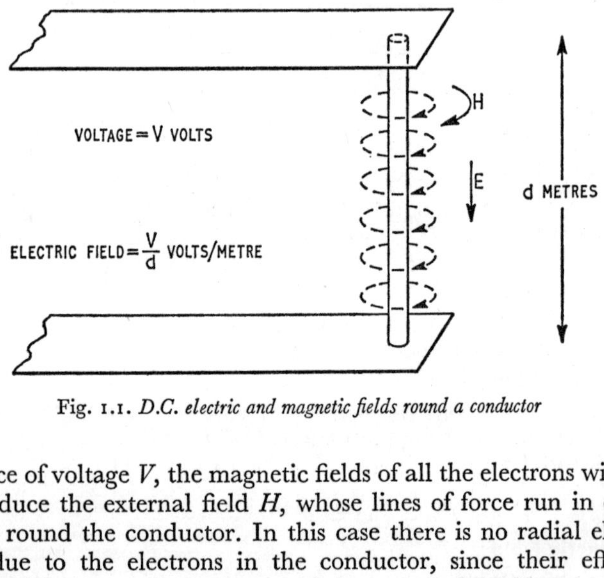

Fig. 1.1. *D.C. electric and magnetic fields round a conductor*

a source of voltage *V*, the magnetic fields of all the electrons will add to produce the external field *H*, whose lines of force run in closed circles round the conductor. In this case there is no radial electric field due to the electrons in the conductor, since their effect is equalized by the nuclear charges of the atoms of the bar.

As with an electric field, so, too, with a magnetic field, there is stored energy. It takes some work to put electrons into motion and so set up a magnetic field, and if a current is increased this is equivalent to adding another magnetic field in the same direction as that already there. The work necessary to do this is illustrated by the force needed to bring two similar magnets together so that their fields add.

Chapter 2

THE MOTION OF ELECTRONS IN STEADY ELECTRIC AND MAGNETIC FIELDS

2.1 Electrons in Steady Electric Fields

In the previous chapter it has been described how electrons are always surrounded by their electric fields, and, if they are moving, have also an encircling magnetic field. Electrons are also affected by external magnetic or electric fields in which they happen to find themselves, and their usefulness in valves arises from the ways in which they are acted upon by such fields.

It is convenient to consider the various cases that can arise in increasing order of complexity, and the first and most simple of these is the effect of a steady electric field. An electric field, by definition, is something which exerts a force on an electric charge, so the charged electron in an electric field will experience a force. When we use the M.K.S. system of units, and measure the electron charge as e coulombs, remembering that it is a negative charge, and suppose that it is placed in a field of E volts per metre, the force on it will be Ee newtons. This will give it an acceleration of Ee/m m/s^2, where m is its mass in kilogrammes. Regarding the direction of this force, either we can remember that the convention for the direction of a field is from positive to negative charge, and that the electron carries a negative charge, so that the force on an electron is always in the opposite direction to that of the field, or it may be remembered as a result of experience that valve anodes are always connected to positive h.t., so electrons always go from negative to positive. As the electron is accelerated by the electric field, the field does work on it, and again as in the case of particles falling under the action of gravity, we often refer to electrons 'falling through' a potential difference. If an electron starting from rest falls through a voltage of V volts, we have already seen that the work done on it, and therefore its increase in kinetic energy, is eV joules. This must equal $\frac{1}{2}mv^2$ where v is its final velocity, which is therefore given by

$$v = \left(\frac{2eV}{m}\right)^{\frac{1}{2}} \text{ m/s} \tag{2.1}$$

or writing $\eta = e/m$ as we shall do in later chapters,

$$v = (2\eta V)^{\frac{1}{2}} \tag{2.2}$$

Putting in some figures, we find that electrons accelerated by only 1 volt are moving at nearly 600 km/s, roughly 370 miles *per second.* If the voltage is raised to 10 kV, quite a normal voltage in transmitting and microwave valves, the speed has increased to about 60,000 km/s; one-fifth of the velocity of light! At higher voltages still, the effect of relativity steadily increases the mass according to the well-known expression

$$m = \frac{m_0}{\left(1 - \frac{v^2}{\mathbf{c}^2}\right)^{\frac{1}{2}}} \tag{2.3}$$

where m_0 = mass when at rest

$\mathbf{c}$ = velocity of light,

while the velocity is given by the expression

$$v = \mathbf{c}\,[1 - (1 + 1{\cdot}96 \times 10^{-6}V)^{-2}]^{\frac{1}{2}} \tag{2.4}$$

where $\mathbf{c}$ is the velocity of light (300,000 km/s). At 300 kV, which is the sort of voltage used in the highest power microwave valves, this gives the electron speed as 230,000 km/s.

2.2 Moving Electrons in Steady Magnetic Fields

Next, let us consider how electrons behave when there is only a magnetic field present. Just as electrons only produce a magnetic field when they are moving, so they are only affected by a constant external magnetic field when moving through it, and only by the component of the field at right-angles to their direction of motion. The field produces a force on the electron, which is proportional to the velocity, but is always at right-angles to the direction in which the electron is going. When the electrons are flowing along a wire, this force is, of course, the well-known motor force, the direction of which is given by the 'left hand rule'. Because the magnetic force always remains at right-angles to the direction of motion, it never does any work, and so can only affect the direction in which the electron is moving; it cannot give the electron any increase of energy. The magnitude of the force is proportional to the flux density B of the magnetic field, to the electron charge e, and to its velocity v at right-angles to the magnetic field. It is therefore given by

$$\text{Force} = Bev \tag{2.5}$$

Any particle being constantly deflected at right-angles to its direction of motion moves in a circular path, the simplest example of this being a weight swinging round on the end of a piece of string when the weight is constantly deflected from its direction of motion at any instant by the tension in the string. By equating the deflecting force Bev to the well-known expression for the force causing acceleration towards the centre, mv^2/r, we find the radius of the circular orbit is

$$r = \frac{mv}{Be} \tag{2.6}$$

Its angular frequency, i.e. the number of radians turned through in 1 second (or 2π times the number of revolutions per second), is usually denoted by ω_c and is given by

$$\omega_c = \frac{v}{r} = B\frac{e}{m} = \eta B \tag{2.7}$$

This is usually called the *cyclotron frequency*, from its use in electron accelerating machines.

It will be noted that while the radius of the orbit depends on the initial velocity v, the cyclotron frequency is independent of v and varies only with the magnetic flux density. The motion is indicated in Fig. 2.1.

2.3 Focusing Action of a Magnetic Field

If the magnetic field is parallel to the direction of motion, it does not affect the electron, and since any magnetic field in any other direction can always be resolved into two components, one in the direction of motion, and the other at right-angles to it, all cases are

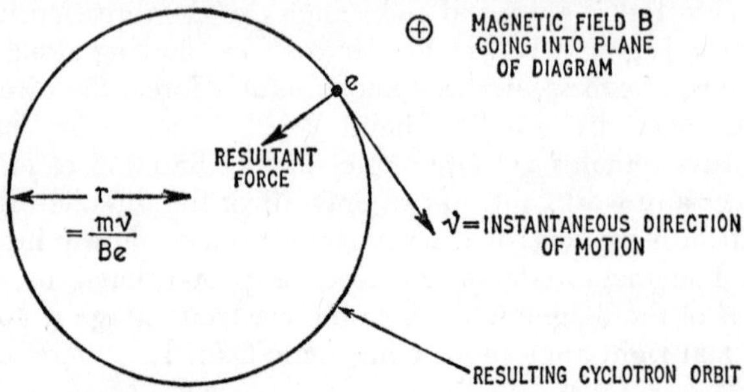

Fig. 2.1. *Cyclotron motion of electron in magnetic field*

covered. One important and useful application of this is in focusing electron streams. To see how this happens, consider an electron stream in which the electrons are diverging at equal speeds from a point on the axis of motion, as shown in Fig. 2.2, such as would be obtained if a beam were incident upon a small hole in a screen. Any electron has its velocity directed mainly parallel to the axis, but with a relatively small radial component which causes the beam to spread. Now let a magnetic field parallel to the axis be switched on. The forward motion of the electron, which is parallel to the magnetic field, is not affected, but the radial straight motion is turned into a circular motion, as shown in Fig. 2.2. All electrons will complete one circle in the *cyclotron* period, $2\pi/\omega_c$, and so they will all again be crossing the axis every $2\pi/\omega_c$ seconds after leaving it. If their axial velocity is v, they will thus be brought to a series of foci at distances

$$\frac{2\pi v}{\omega_c}, \frac{4\pi v}{\omega_c}, \frac{6\pi v}{\omega_c} \text{ etc.}$$

along the axis.

2.4 Parallel Electric and Magnetic Fields

The next case is that of both electric (E) and magnetic (B) fields together, and in the same direction. This is a simple one, since if the electrons are initially moving parallel to the fields they are merely accelerated by the E field, as in the first case considered, without being affected by the B field, while if they are moving with a component at right-angles to the fields this component is turned into cyclotron motion, i.e. they execute circles at a constant frequency, whose radius depends on the initial velocity. It will be apparent that the focusing action will still take place, but since the velocity will be increasing due to acceleration of the E field, the successive foci will get farther and farther apart. Before leaving this case, we should look again at the focusing action already described. It can be seen that if, as described, the electrons are diverging from a point on the axis, by considering the movement of each one of them they will at first diverge as they follow their individual helical paths and will be brought together at the first focus. The resultant electron stream will then have a 'scalloped' character, as shown. Here we have considered the action of the external fields, but have ignored the forces exerted by the electrons on each other—the space-charge forces. We shall return to these later, but we may note here that even when

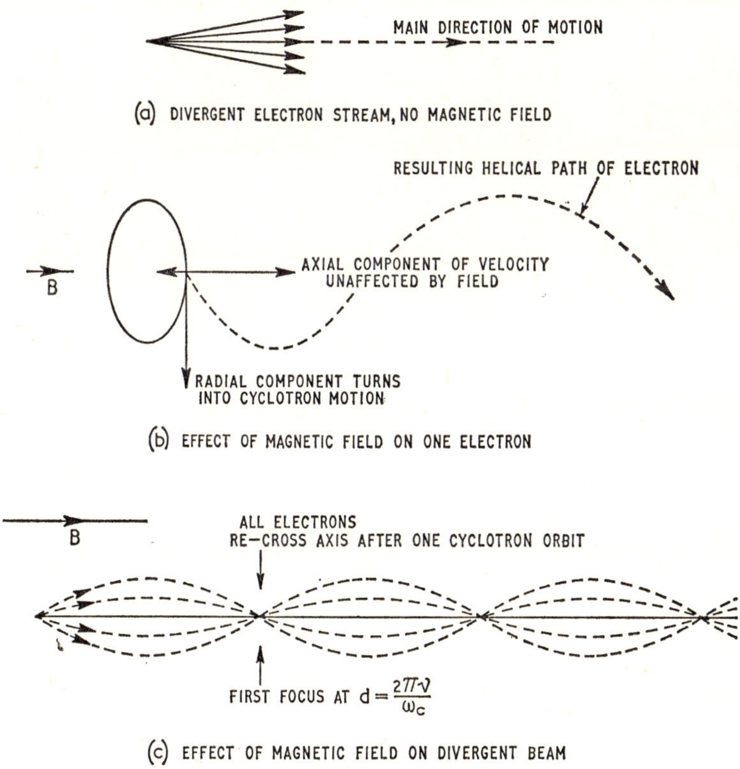

Fig. 2.2. *Focusing action of magnetic field on thin electron beam*

these additional forces are taken into account it is still possible to focus dense electron beams by placing them in an axial magnetic field. This property is extensively used in microwave valves where it is necessary to make electron beams hold together over long distances.

2.5 Crossed Electric and Magnetic Fields

The last, and most complex, case is when there are both E and B fields present together, at right-angles to each other. They are then referred to as *crossed-fields*, and are, as will be seen, of great interest and use in an important family of valves. A simple illustration of this case is as shown in Fig. 2.3.

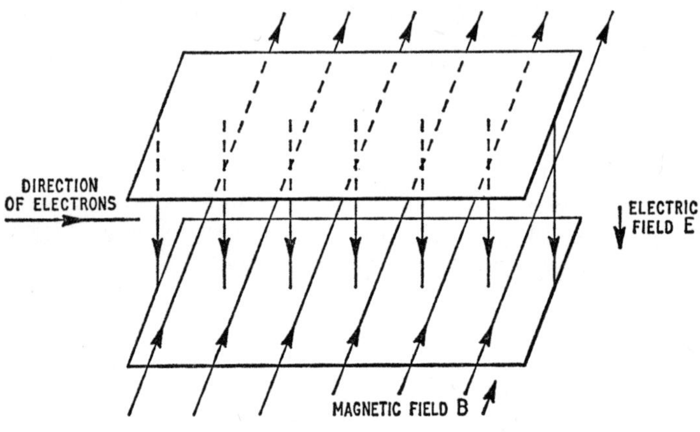

Fig. 2.3. *Crossed-field arrangement*

What electrons do when they encounter such a system depends on how they enter it. For instance, it can be seen that if they are going in the direction shown, at right-angles to the fields, at that particular speed $v = E/B$ which makes the downward magnetic deflecting force (Bev) just equal to the upward electric deflecting force (Ee), the electrons will continue to move unperturbed. This is the simplest case, but it has one result that is common to all cases, namely that the mean velocity in the forward direction, at right-angles to both E and B fields, is E/B.

Another case, slightly more difficult, is that of electrons starting from rest from the lower plate, which would be approximately true if a cathode were placed there. At first the electrons would only be affected by the electric field, and would start to move towards the opposite plate. As soon as they move, however, they interact with the magnetic field, and will be deflected sideways. Their velocity continues to increase as they pass through the electric field, until finally they are moving at a speed $2E/B$, parallel to the plates. They continue to be deflected by the magnetic field, which now makes them move against the electric force, so that they slow down, and finally come to rest just as they again reach the lower plate. They have taken one cyclotron period, $2\pi/\omega_c$, to do this, and have advanced a distance $(2\pi/\omega_c)$ (E/B) in the direction parallel to the plates, i.e. as if they had been moving uniformly at a speed E/B. Their motion, which is cycloidal, is accurately thought of as being a uniform motion of speed E/B added to a circular motion at the cyclotron angular frequency. They have moved just as if they were points on the rim of a wheel rolling along the bottom plate at a speed of E/B. Returning to the case of unperturbed motion, it can be seen that those electrons moved as if they were points at the centre of such a wheel, whose speed remains constant. This concept is valuable also for many other cases, and the motion of any electrons injected at speeds of up to $2E/B$ can be described by the motion of a suitable point somewhere inside our rolling circle.

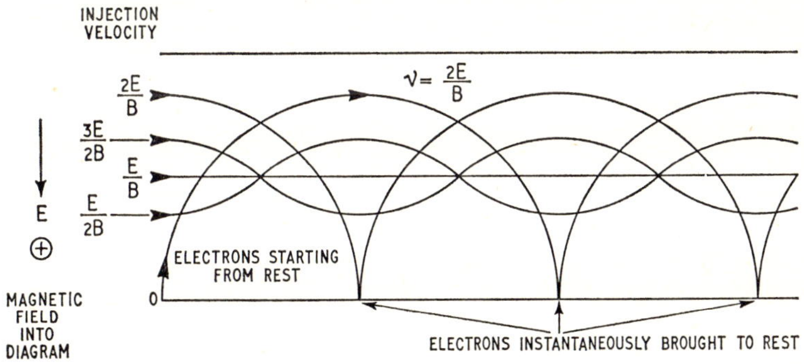

Fig. 2.4. *Electron trajectories in crossed-fields*

Some of the possible trajectories are shown in Fig. 2.4 and by using the idea of a rolling wheel, the reader will readily be able to draw any others. Injection at speeds greater than $2E/B$ is not normally encountered. The trajectories are somewhat similar, but

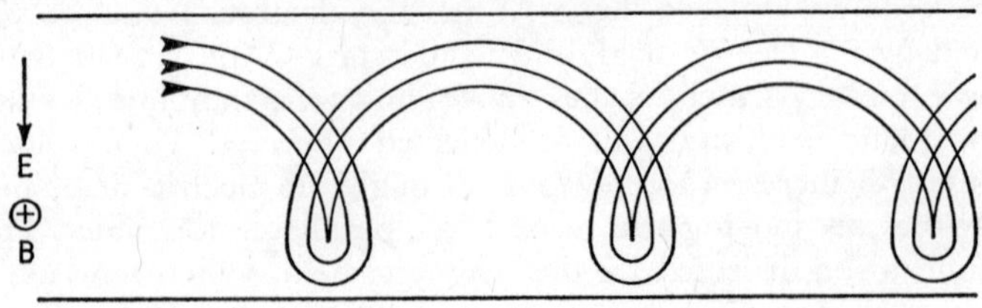

Fig. 2.5. *Cycloidal orbits for injection velocities greater than* $\frac{2E}{B}$

now contain loops, as shown in Fig. 2.5; at the lowest points of these, the electrons are actually going backward. For all these trajectories, however, the mean speed over any complete number of cycloids is E/B, and its direction is always at right-angles to both the electric and magnetic fields.

BIBLIOGRAPHY

KLEMPERER, O., *Electron Optics*, Cambridge University Press (1953).
HARMAN, W. W., *Fundamentals of Electron Motion*, McGraw Hill (1953).

Chapter 3

PROPERTIES OF R.F. CIRCUITS

3.1 Alternating Fields

So far we have only considered the production of steady (i.e. non-time-varying) fields, and the motion of electrons in them. The most general purpose of thermionic valves is, however, for the generation or amplification of alternating power, and in this use the electrons are acted upon by alternating (i.e. time-varying) electric fields, in addition to any steady fields which may be present. The various interaction processes between the electrons and alternating fields are discussed in subsequent chapters; to understand these it is first desirable to describe the circuit arrangements by which the necessary fields are produced in the regions where they are required.

The reader will be familiar with alternating voltages at mains frequencies, and that these are of sinusoidal form having a variation with time given by

$$V = V_0 \sin \omega t \qquad (3.1)$$

where V_0 is the peak value of the voltage and $\omega = 2\pi f$, where f is the frequency in cycles per second, when t is measured in seconds. The alternating voltages, currents and fields we shall be concerned with, right up to the microwave region, have this same sinusoidal form, with the appropriate value of ω.

3.2 Field, Voltage and Current in a Plane Capacitor

The simplest representation of the part of a valve between any two electrodes is that of a capacitor, in which the electric fields are perpendicular to the surfaces. The plates of the capacitor may be plane parallel surfaces or concentric cylinders, but for ease of discussion we shall assume they are plane parallel. In order to produce an alternating field across such a capacitor we may connect it to an a.c. supply of the required frequency, which, for uniformity with the cases that come later, we shall take as a current source $I_0 \sin \omega t$, as shown in Fig. 3.1. As the source current varies, electrons are pumped

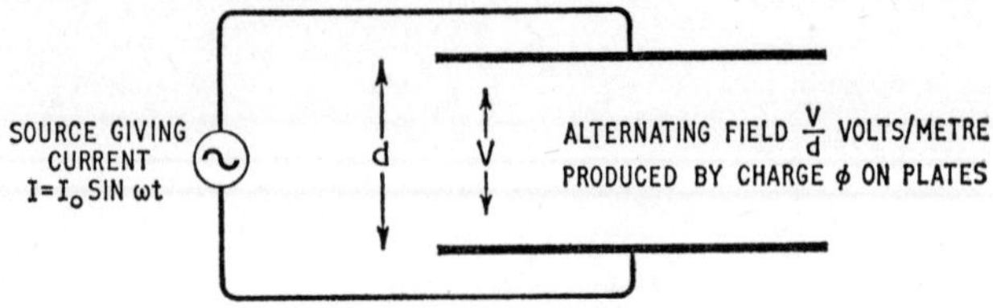

Fig. 3.1. *Field in a plane capacitor*

round the circuit to vary the charge on the capacitor, and therefore the voltage between its plates, which is proportional to the charge. As the current into a capacitor is the rate of increase of charge, then the rate of increase of voltage is proportional to the current. Hence, at the instant when the current is zero the voltage is at a maximum, and when the current has its maximum value the voltage is zero. This situation, shown in Fig. 3.2, is the familiar one that, in a capacitor, the voltage and the current are in quadrature, i.e. $\pi/2$ rad (90°) out of phase, with the current leading the voltage. If the value

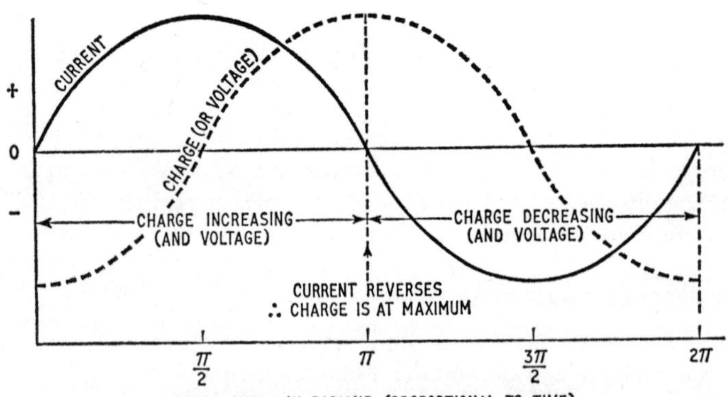

Fig. 3.2. *Current, charge and voltage in a capacitor*

of the capacitance is C farads, which has an impedance of $1/\omega C$ ohms, then the voltage will be given by

$$V = \frac{I_o}{\omega C} \cos \omega t \text{ V} \tag{3.2}$$

This voltage produces a field between the plates given by

$$E = \frac{V}{d} = \frac{I_o}{\omega C d} \cos \omega t \text{ V/m} \tag{3.3}$$

where d is the spacing between the plates in metres.

3.3 Production of Fields between Valve Electrodes

It may be noted that this field is only relatively slightly changed if, instead of a continuous plate, a wire grid is substituted for either electrode, provided that the grid wire spacing is considerably less than the spacing between the two electrodes (Fig. 3.3).

Fig. 3.3. *Effect of replacing capacitor plate by grid*

It will be seen that, because ω appears in the denominator of Equation 3.3, the field across the capacitor decreases with frequency, when the current is kept constant. It is usual therefore to connect an external impedance in parallel with the valve capacitance, so that the voltage developed is determined by this external impedance rather than by the valve capacitance alone. There would be little object in making this external impedance a capacitance, since this would only lower the voltage developed, and the variation of field with frequency would not be changed.

At relatively low frequencies, say up to 100 kc/s, it is common to make the external impedance a resistance whose value is low compared with the impedance of the capacitor. With conventional valves (triodes, pentodes, etc.) the capacitance between electrodes is small, of the order of 10 pF, and the impedance of this at 100 kc/s is about 160,000 ohms. A suitable external resistance might then be 10,000 to 20,000 ohms. For example, in the circuit of Fig. 3.4, where

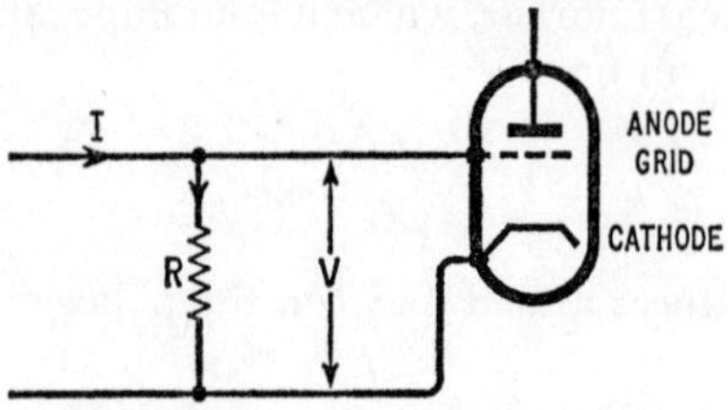

Fig. 3.4. *Grid-cathode field produced across resistance in triode valve*

we meet for the first time the conventional circuit symbol for the three-electrode valve or *triode*, the field between the centre electrode, the grid, and the lower electrode, the cathode, is due to the voltage V developed across the resistance R by the current I flowing in it. Such a circuit is useful for working over a wide frequency band, since the value of the resistance and therefore the field produced is independent of frequency. In this simple diagram we have omitted any d.c. power supplies, and it must be remembered constantly that only alternating quantities are being considered.

A third possibility for use as the external impedance is an inductor. If this has an inductance of L henries, and therefore an impedance of ωL ohms at a frequency of $\omega/2\pi$ cycles per second, an alternating voltage of peak value $I_0\omega L$ volts will be developed across it. This voltage will again be in quadrature with the current, but in this case the voltage leads the current, and if drawn in Fig. 3.2 would be 180° out of phase with the capacitive voltage shown there. It is also important to note that the voltage produced, and therefore the field in the valve, increases with frequency.

Inductance is, of course, associated with changing magnetic flux, so that, under the above conditions, there will be an alternating magnetic field due to the inductor. In general there must always be an alternating magnetic field associated with an alternating electric field. Electrons in motion, however, do not interact with the alternating magnetic fields until the electrons are moving at relativistic speeds (say, at voltages from 30 kV upwards), so that this interaction will be ignored.

3.4 Resonant Circuits

Now consider a circuit made up of capacitance C, inductance L and resistance R in parallel, as in Fig. 3.5. It was seen above that increase of frequency caused an upward change in the impedance of

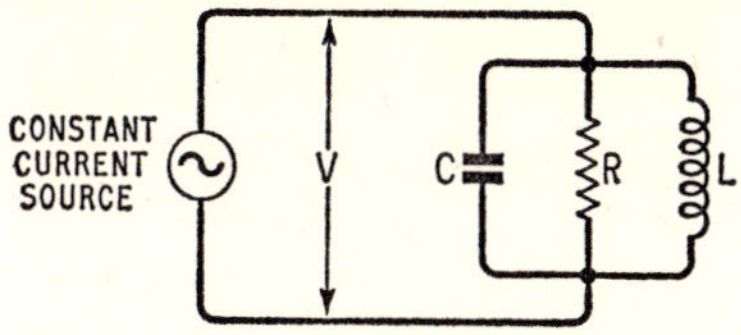

Fig. 3.5. *Parallel resonant circuit*

the inductor and a downward change in the impedance of the capacitor. There must therefore be a unique frequency at which the two impedances are equal, given by

$$\omega_0 L = \frac{1}{\omega_0 C} \text{ or } \omega_0^2 LC = 1 \qquad (3.4)$$

The currents in L and C will then be equal in magnitude and exactly opposite in phase, so that the current flowing into one is supplied by the current flowing out of the other, and the externally supplied current I only flows in the resistance R, producing a voltage IR. This unique frequency is called the resonant frequency of the circuit. For a given capacitance C and current I the effect of 'tuning' to resonance by adding the inductance L, is to increase the current through the capacitance, and therefore the voltage across it. The ratio of the increase is called the Q or magnification factor, and is given by

$$Q = \omega_0 CR = R\sqrt{\left(\frac{C}{L}\right)} \qquad (3.5)$$

At frequencies on either side of the resonant frequency the currents in the capacitor and inductor will no longer balance out, and the difference will have to come from the supply. There will then be less current in R, and the voltage developed across the circuit will be less. The curve relating the voltage developed at constant current to the frequency is called the resonance curve, a typical curve being shown in Fig. 3.6. When the width or flatness of this curve is greater, the circuit is said to have a wider bandwidth. This bandwidth is usually specified in terms of the frequency difference between the points where the power in the circuit has dropped to half that at the peak or, since the power is proportional to the square of the voltage, when the voltage has dropped to $1/\sqrt{2}$ of that at the peak. It may easily be shown that this bandwidth, $\delta\omega$, is given by

$$\delta\omega = \frac{1}{CR} = \frac{\omega_0}{Q} \qquad (3.6)$$

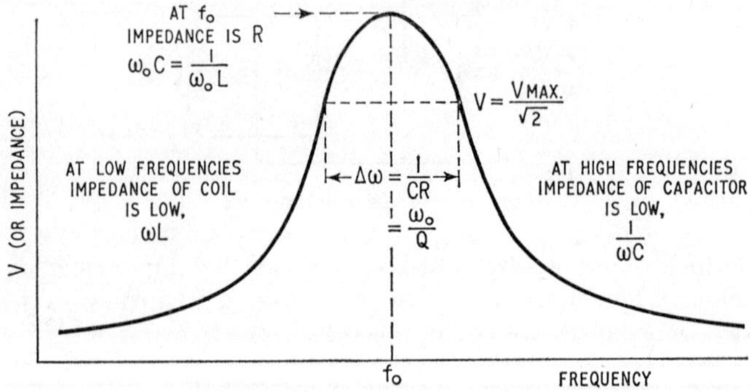

Fig. 3.6. *Impedance of resonant circuit*

Such circuits can, in principle, be used from the lowest frequencies, and are met with in practice from about 10 kc/s up to a few hundred Mc/s. At the low frequencies the resistance is usually the unavoidable loss resistance of the inductor. At broadcast frequencies *Q*'s of 100 to 200 are common, giving bandwidths of the order of 10 kc/s.

In the form the tuned circuit has been drawn in Fig. 3.5 it must be remembered that a low loss coil has a high parallel resistance and conversely high loss will be given by low resistance. For the wide band circuits used in television receivers actual resistance usually has to be connected in parallel with the inductor to increase the bandwidth.

3.4.1 A Resonant Cavity as the Limit of an LC Circuit

As the operating frequency is increased the coils of the inductor and the value of the external capacitance must be reduced to maintain resonance, until eventually the circuit assumes the form of Fig. 3.7(a), with a small single-turn coil tuned only by the capacitance between the valve electrodes.

The losses in such a coil are high in relation to its inductance, since the wire resistance is increased at the higher frequencies by the 'skin effect', i.e. the tendency of the current or movement of charge to be confined closely to the surfaces of conductors. The loss of the coil can be greatly decreased, and its parallel resistance correspondingly increased by increasing the surface area of the conductor. This area can be increased to the greatest extent, while keeping the same resonant frequency, by making it as shown in Fig. 3.7(b), or ultimately by extending it right around the valve as in Fig. 3.7(c). The circuit has become a *cavity resonator*, which has the properties that the current is now confined to the inner surface of the conductor, and the alternating magnetic field is completely enclosed. Hence, unlike the coil of Fig. 3.7(a), the resonator cannot act as an aerial, and no power is lost by radiation. The magnification factor Q which can be obtained from such a cavity is accordingly very high, e.g. about 4,000 at a frequency of 3,000 Mc/s.

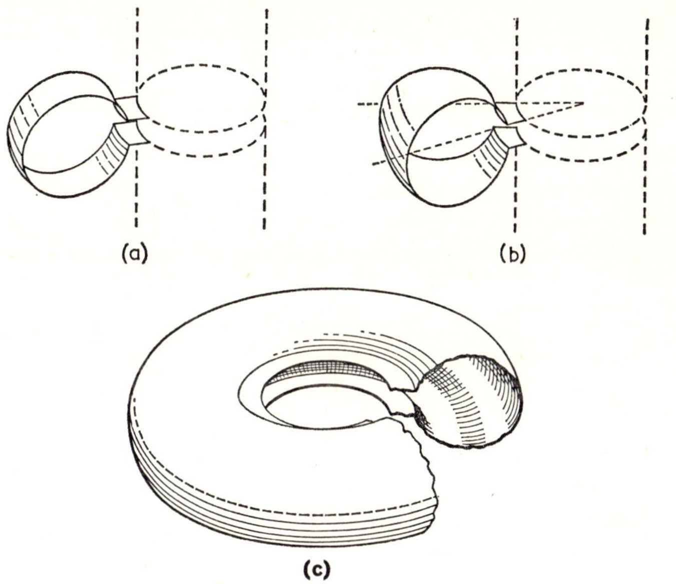

Fig. 3.7. (*a*) *Single loop coil.* (*b*) *Reduction of loss.* (*c*) *Resonant cavity*

3.5 Transmission Lines at Radio Frequencies

Turning now to a different way of generating fields, which will also in time lead us back to resonant cavities, let us consider *transmission lines*. A transmission line is just the system of wires or conductors, or even dielectrics, by means of which power is transported from one place to another. Perhaps the simplest model is the co-axial line, the use of which is familiar as a television aerial lead, or microphone or pick-up lead. It consists of an inside and an outside conductor, the space between them being filled with dielectric, commonly polythene. The transport of power along the line is by *electromagnetic waves*. In free space these waves travel or propagate at the speed of light i.e. 3×10^8 m/s, which is usually denoted by the symbol **c**. In the coaxial line they propagate more slowly, at the speed of light through the dielectric, which is given by

$$v = \mathbf{c}/(\mu\epsilon)^{\frac{1}{2}} \tag{3.7}$$

where μ and ϵ are the permeability and permittivity of the dielectric relative to that of free space. For polythene v is approximately $2\mathbf{c}/3$. Since the wavelength and frequency f of a wave are related by the expression

$$v = f\lambda \tag{3.8}$$

then, for the same frequency, the wavelength in the coaxial line is about two-thirds of the free-space wavelength.

3.6 Standing Waves on Transmission Lines

At microwave frequencies the line will often be many wavelengths long, and the uniform pattern in it (Fig. 3.8) which occurs at

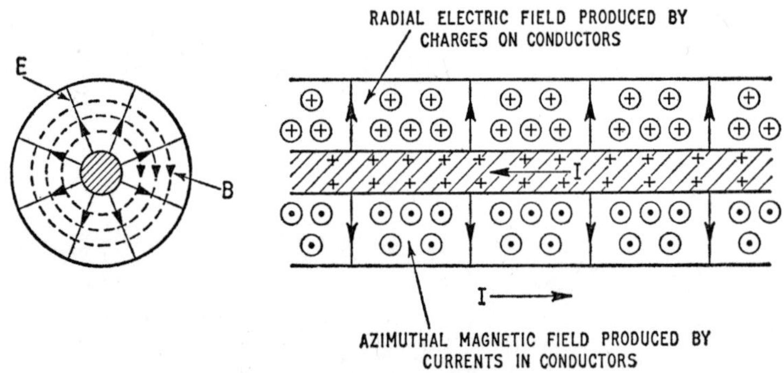

Fig. 3.8. *Coaxial cable with d.c. fields*

d.c. and low frequencies will be changed as shown in Fig. 3.9. The electric charge on the conductors, like the electric and magnetic fields between them, will then vary sinusoidally along the line length, the complete pattern forming a travelling wave moving along in the direction of propagation at the velocity v. If an observer moved along with the wave at the same velocity, he would detect a constant field whose magnitude would depend on the particular phase at the

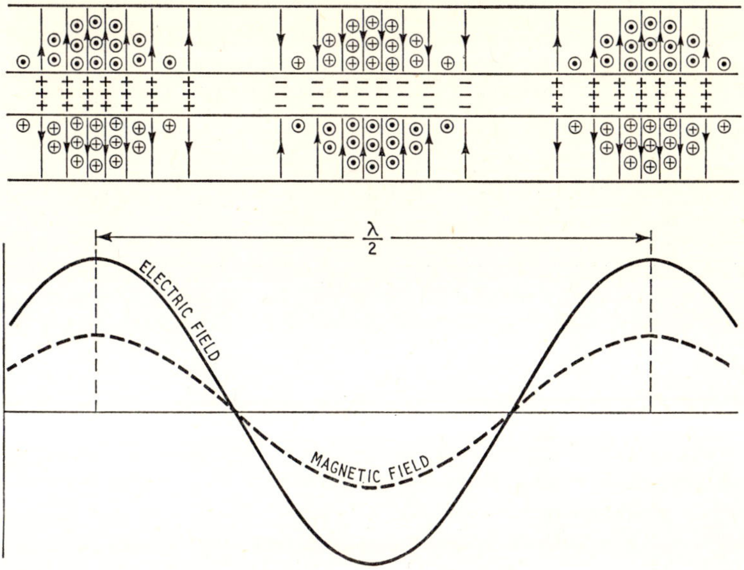

Fig. 3.9. *R.F. field pattern in coaxial cable*

moving point, while if a stationary detector, which measures the time average r.m.s. field, were placed at any point along the line, it would indicate the same value whatever the position. This is true as long as the wave meets nothing which causes any reflections. If reflections do occur, the reflected waves travel back along the line, and then interference between the forward waves and the backward waves causes a *standing wave pattern* to occur. When this happens there exists in the line a stationary pattern of field variation, which can be detected by moving a detector as above, for example a small probe feeding a crystal, along the line as shown in Fig. 3.10. It can be shown that this variation has half the wavelength of the r.f. propagating along the line, and so this behaviour gives one way of measuring wavelengths.

If this behaviour seems unfamiliar to the reader, it may be useful to mention the stretched string analogy. We know that if we take a very long stretched string and shake the end of it we see the waves we excite travelling along the string. If the other end is fixed, then waves are reflected, and we find standing waves along the string. In particular, we can find a number of resonant frequencies, where the

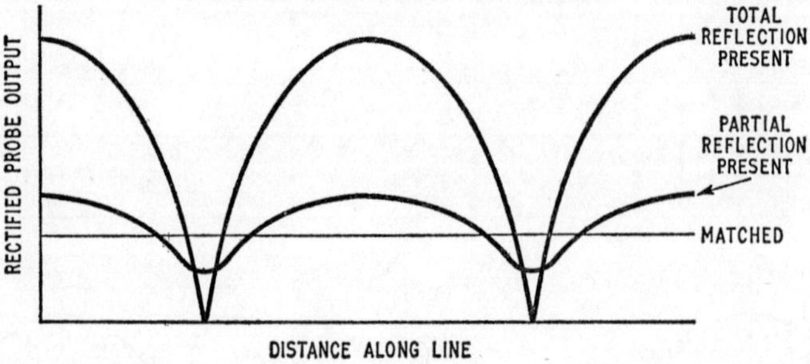

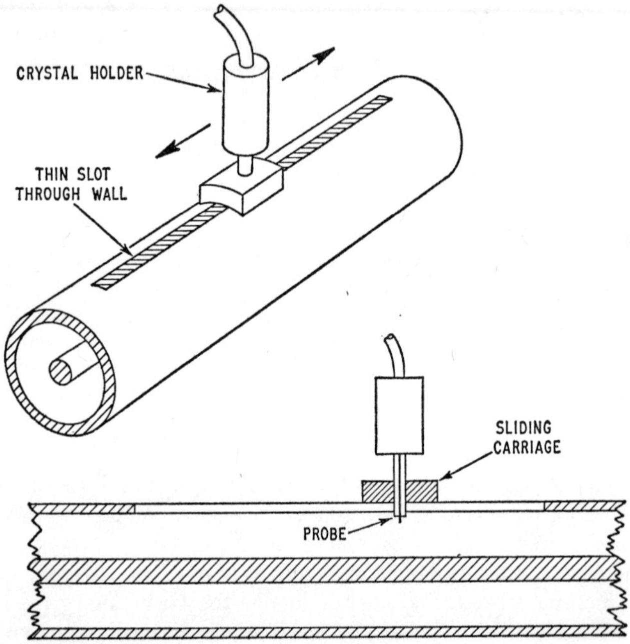

Fig. 3.10. *Sliding probe in slotted line*

string is a whole number of half-wavelengths long, where the standing wave pattern builds up to a large amplitude, giving a long stationary pattern of movement along the string (Fig. 3.11).

This resonant behaviour is also found in transmission lines. If we take a length of transmission line that is short-circuited at both ends,

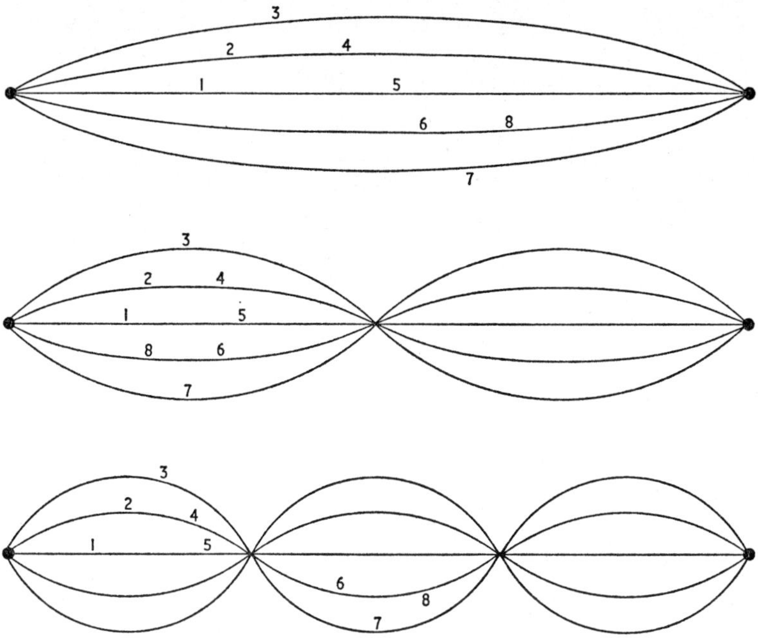

Fig. 3.11. *Standing wave patterns on stretched string. Figures denote position of string at successive instants of $\frac{1}{8}$th period of oscillation*

and excite it by means of a probe at some intermediate point, at a frequency such that the line is just a half-wavelength long, the waves that travel away from the probe in both directions will be reflected to and fro from the short-circuited ends, and will cause a large field to build up in the centre of the line, while large currents will flow round the short circuits at the ends.

This is, of course, similar to the behaviour we saw earlier in the resonant cavity, the description of which was approached by considering the development of a single-turn coil, whose ends are bent to form a capacitor. At the resonant frequency there will be, at alternate quarter periods, peak electric fields across the capacitive part of the circuit and peak magnetic fields round the inductive part of the circuit, as shown in Fig. 3.12. Because the two plates of the capacitor are instantaneously in opposite phase, the length of the wire joining them must be about a half-wavelength long; consequently the diameter of the coil will be about $\lambda/2\pi$. With no dielectric present, $f\lambda = c$, and therefore at 1 Mc/s, $\lambda = 300$ m, while at 1,000 Mc/s, $\lambda = 30$ cm. It is clear therefore why it is only at microwave frequencies that such cavities are used, and that their

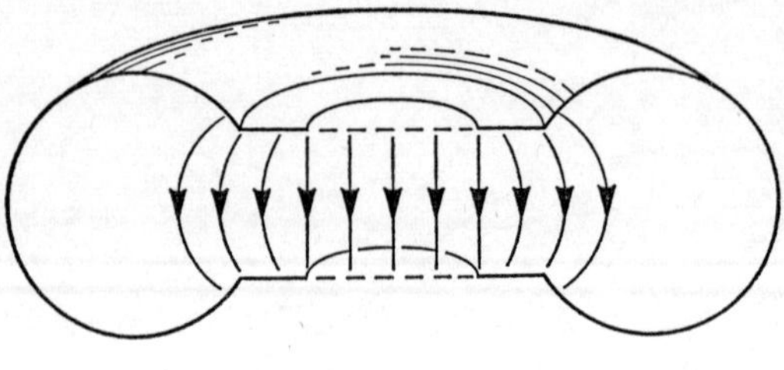

(a) AT $t=t_0$, PEAK CHARGE ACROSS GAP CAUSES PEAK E FIELDS

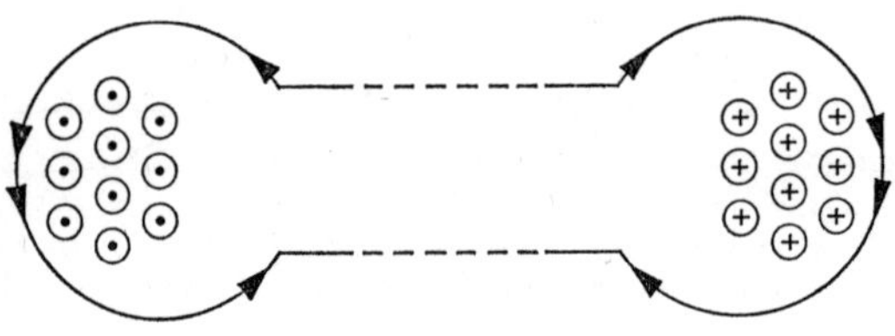

(b) AT $t=t_0+\frac{T}{2}$, PEAK CURRENT FLOW IN WALLS CAUSES PEAK B FIELDS

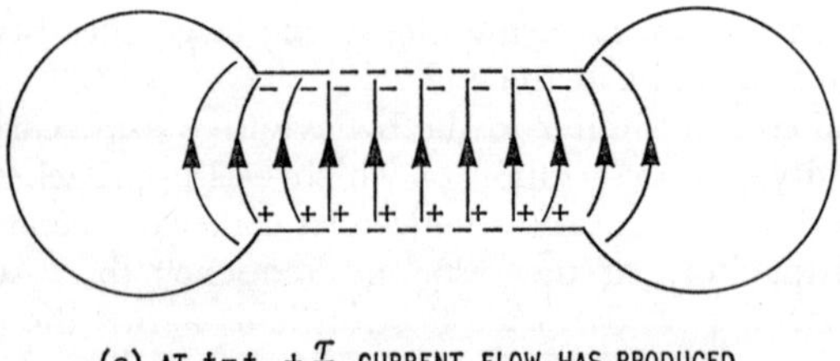

(c) AT $t=t_0+\frac{T}{2}$, CURRENT FLOW HAS PRODUCED CHARGE BUILD-UP ACROSS GAP OF OPPOSITE SIGN

Fig. 3.12. *Fields and current in resonant cavity at successive quarter-periods* use can extend only up to about 300,000 Mc/s (λ = 1 mm), when they become impossibly small to make.

3.7 Characteristic Impedance and Matching

If a uniform transmission line were infinite in length, no reflections would occur on it, since a wave launched from one end would never meet any discontinuity. Suppose we take a finite length of transmission line; then we can terminate it in an impedance Z_o so that no reflections occur. This terminating impedance must look to the wave like another piece of transmission line, and is therefore called the *characteristic impedance* of the line. A line which is terminated in its characteristic impedance is said to be matched, and since no reflections return from the matched termination, this is clearly the condition that all the power flowing along the line is delivered into the termination. A matched line will present its characteristic impedance at whatever point it is measured, since it always looks like an infinitely long line.

3.8 Impedance Change along Short-Circuited Lines

In the short-circuited line considered above, however, the conditions are quite different; at its ends there is a maximum current flow, but no field, and so the impedance at these points is zero, while at its centre, a quarter of a wavelength away, there is a maximum field but no current, which corresponds to an infinite impedance. The impedance at any intermediate point is given by

$$Z = jZ_o \tan \frac{2\pi x}{\lambda} \qquad (3{\cdot}9)$$

where x is the distance from the short circuit. This variation is shown in Fig. 3.13.

Here is another way therefore of feeding power from a source to develop high electric fields. If a section of coaxial line a quarter-wavelength long is short circuited at one end, and fed at a suitable point from a small probe or loop coupling, it will resonate and develop large fields across its open end, which may be connected to the electrodes of a valve, which is usually arranged so that it will plug into the coaxial arrangement, as shown in Fig. 3.14. Although this may appear to be a different arrangement from the resonant cavity described previously, the difference is more in the way we have thought about them than between the systems themselves, as can be seen when they are compared side by side, as in Fig. 3.15.

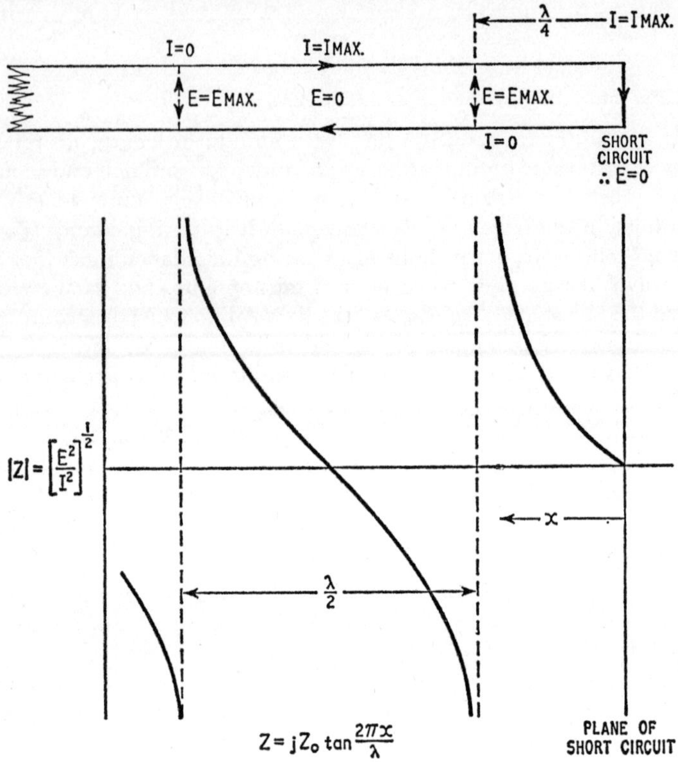

Fig. 3.13. *Impedance transformation along short-circuited line*

3.9 Dispersion, Attenuation and Propagation Constants

Transmission lines can be of other forms than coaxial, and twin wires or parallel plates may be used. Like the coaxial line, these both have two conductors, and consequently, when infinitely long or terminated in their characteristic impedance, have no low-frequency cut-off; that is, they would be of use equally for d.c., 50 c/s mains, or 100 Mc/s r.f. power. Lines like this which behave the same for all frequencies are said to be *non-dispersive*. One particular property is that the velocity with which the wave peaks travel along the line, which we previously called the velocity of the wave, but which is more properly called the *phase velocity* v_p, is constant for all frequencies. For any circuit, this velocity is defined by the product of

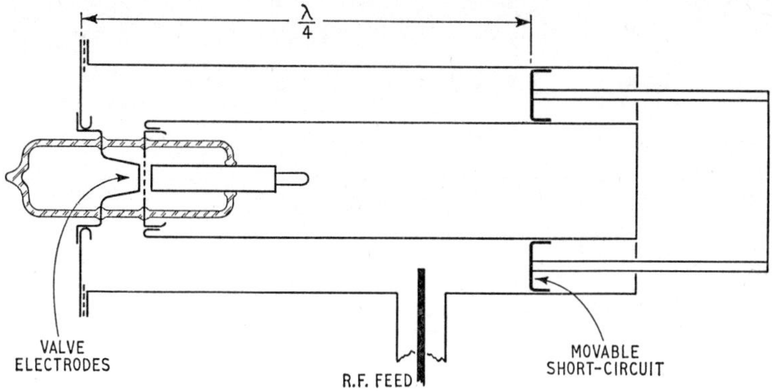

Fig. 3.14. *Coaxial-line resonator for use with valve*

the frequency and the wavelength, i.e. Equation 3.8 should really be

$$v_p = f\lambda \tag{3.10}$$

In a non-dispersive system, such as free space, the wavelength therefore varies inversely as the frequency. In dealing with microwave structures and transmission lines it is often more convenient to use the angular frequency ω ($= 2\pi f$) rather than the frequency itself, and a quantity β ($= 2\pi/\lambda$) instead of the wavelength. We then have

$$v_p = f\lambda = \frac{\omega}{\beta} \tag{3.11}$$

where β is called the propagation constant. The angular frequency is, of course, the number of radians of the wave that pass a fixed observer in unit time (1 sec). Correspondingly, β is the number of radians of the wave that instantaneously occur in unit length. If, therefore, the observer moves a distance z in time t in the direction of the wave motion, the number of radians that pass him will be $\omega t - \beta z$. The voltages, fields and currents in the wave will therefore be described by the real or imaginary parts of expj $(\omega t - \beta z)$ i.e. cos or sin $(\omega t - \beta z)$.

One advantage of using the exponential form expj $(\omega t - \beta z)$ is that loss can readily be taken into account in the same form. When the propagation is lossless, i.e. the wave does not change in amplitude along the transmission line, as is usually taken to be the case with short lines, then β is real. If, however, there is loss, the wave amplitude falls, or is attenuated, with distance. For an attenuation $e^{-\alpha}$

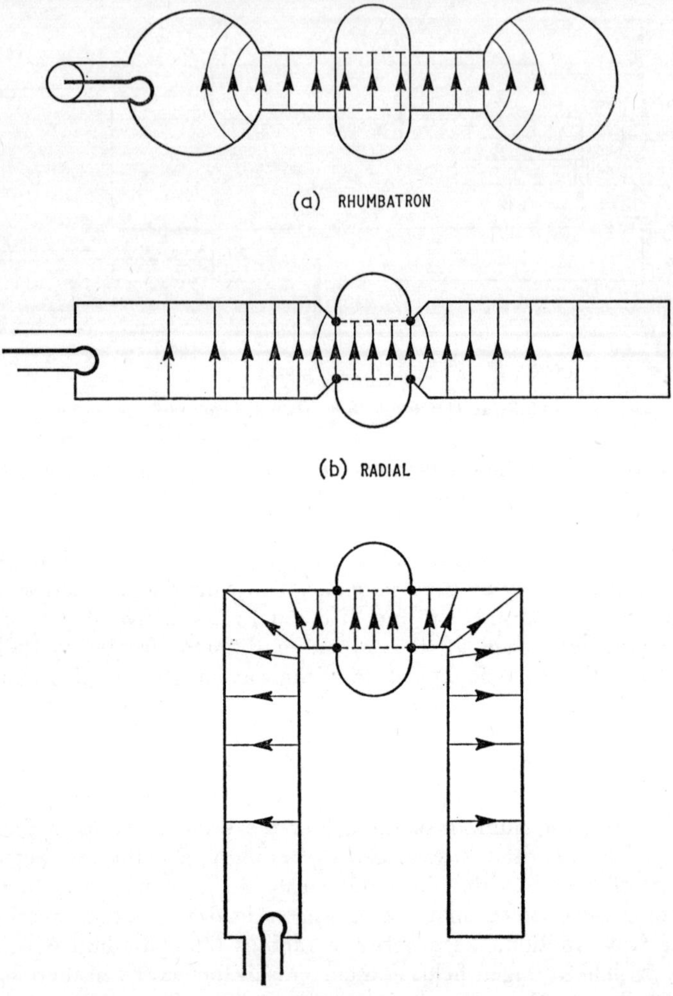

Fig. 3.15. *Forms of resonant cavity: (a) rhumbatron, (b) radial, (c) coaxial*

($\doteqdot 1 - \alpha$ when α is small) per unit length, the total attenuation for a length z will be $e^{-\alpha z}$. The voltages, etc., will then be described by

$$e^{-\alpha z} e^{j(\omega t - \beta z)} = e^{j(\omega t - \beta' z)} \tag{3.12}$$

where $\beta' = \beta - j\alpha$, i.e. the propagation constant has an imaginary component.

3.10 The ω-β Diagram

It will be seen that, using these new variables, ω will be directly proportional to β for a non-dispersive line, so that if one is plotted against the other, to give what is known as an $\omega - \beta$ diagram, as in Fig. 3.16, the graph obtained is a straight line, passing through the

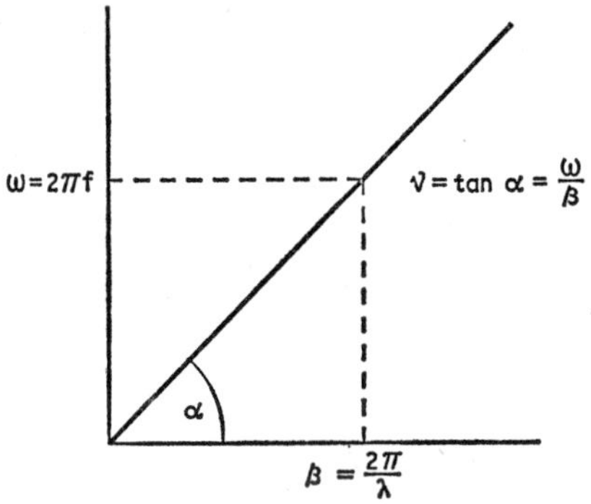

Fig. 3.16. *Propagation on a non-dispersive line*
$v_{phase} = v_{group} =$ *constant*

origin, whose slope is equal to the phase velocity. For dispersive circuits, in which category fall most of the circuits used in microwave valves, the diagram, which is one form of dispersion curve, usually consists of curved lines. The phase velocity at any point, however, is still given by ω/β, that is, the slope of the line joining the point to the origin.

3.11 Phase and Group Velocity

It can fairly easily be shown that the phase velocity is not the velocity with which energy or intelligence is propagated. A continuous sinusoidal signal by itself cannot convey intelligence. It is only when it is stopped, started or otherwise altered (modulated) that intelligence is conveyed, i.e. when other frequencies are involved. These other frequencies usually lie in a small range (called the side-bands) about the original frequency, the sum of the individual sinusoidal components making up the modulated waveform. On a dispersive line these individual waves will shift in phase relative to one another as they travel down the line. If the dispersion is not too great, the effect can be seen to be an alternating increase and decrease or modulation of the wave amplitude. This modulation moves along, not at the same velocity as the wave peaks, but at a new velocity called the group velocity which is faster or slower than the phase velocity depending on whether the higher or lower frequency components move faster.

Suppose we select any two of the individual waves, with angular frequencies ω and $\omega + \delta\omega$, and corresponding propagation constants β and $\beta + \delta\beta$, these latter being by definition the overall phase changes (in radians) which occur in unit length for each wave. Considering first the waves as fixed, and an observer moving relative to them, then the phase of one wave will change by $\delta\beta$ relative to the phase of the other in unit length. Since a change of 2π gives the spacing between one modulation maximum and the next, the relative phase change $\delta\beta$ must therefore represent for the modulation the same function as do β and $\beta + \delta\beta$ for the individual waves. That is, $\delta\beta$ is the propagation constant of the modulation. Next, considering the waves moving relative to a fixed observer, the rate at which the individual peaks pass are respectively $\frac{\omega}{2\pi}$ and $\frac{\omega + \delta\omega}{2\pi}$ in unit time. During this time the waves will reinforce each other $\omega/2\pi$ times, which is therefore the rate at which the modulation peaks pass the observer. The angular frequency of the modulation is therefore dω. By analogy with Equation 3.10 the velocity of the modulation will therefore be given by

$$v_g = \frac{\delta\omega}{\delta\beta} \tag{3.13}$$

For small dispersion $\delta\beta$ will be proportional to dω, so that exactly the same result would be obtained if a large number of sinusoidal components were considered. This velocity v_g is therefore the group velocity visualised above, at which energy is transmitted. It is given at any point on an ω–β diagram by the slope of the actual dispersion curve at that point, as shown in Fig. 3.17.

3.12 Waveguide and its Dispersion

An example of a dispersive circuit is provided by a hollow waveguide. Unlike the lines considered previously, this is composed of only one conductor. It has been known since the nineteenth century

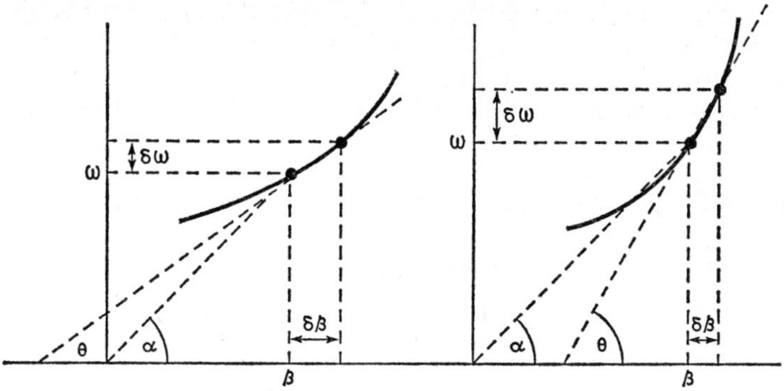

Fig. 3.17. *Dispersive Propagation*

$$v_{phase} = \frac{\omega}{\beta} = \tan \alpha$$

$$v_{group} = \frac{\delta\omega}{\delta\beta} = \tan \theta$$

that r.f. power can propagate along a hollow pipe, provided that the pipe is large enough relative to the wavelength. In the case of rectangular waveguide, the condition for propagation is that the broad dimension of the waveguide is greater than a half-wavelength. This gives a low-frequency cut-off at that frequency where the guide is just a half-wavelength wide; the phase velocity then is infinite and the group velocity zero. As the frequency is increased above this cut-off the phase velocity decreases and the group velocity increases and both approach the velocity of waves in free space. The $\omega - \beta$ diagram of a rectangular waveguide given in Fig. 3.18 shows this behaviour.

3.13 Rectangular Waveguide

One way of looking at propagation in a rectangular waveguide, in the mode usually used, is that of a plane wave progressing along it at an angle to the side walls, and which is reflected alternately from them. Consequently there will be a large standing wave pattern across the guide, the field being necessarily zero at the sides, where there is a conducting wall, and rising to a maximum at the centre (Fig. 3.19). This is equivalent to saying that the field pattern in this mode in rectangular waveguide always arranges itself so that there is a half-wave pattern across the guide. The progression along the guide then arises from reflections at the walls as indicated in

Fig. 3.20. In this diagram the lines *AB* and *CD* represent successive positive and negative wave fronts in a wave travelling along the waveguide from left to right. The free space wave travels at the velocity of light between reflections, at an angle θ to the walls, so that points such as *E* and *G* at opposite edges of the waveguide are

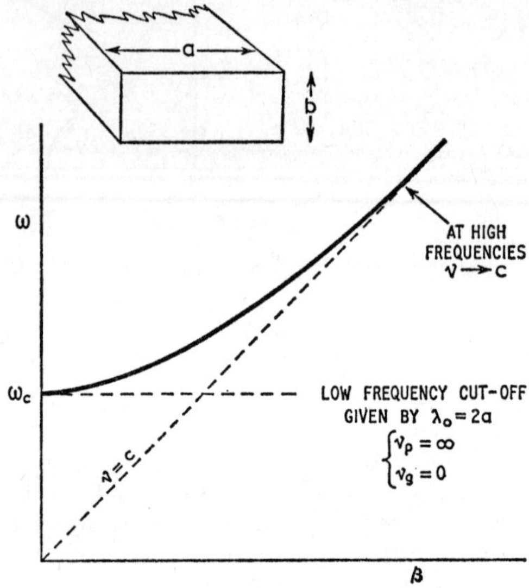

Fig. 3.18. *Propagation in fundamental mode in rectangular waveguide*

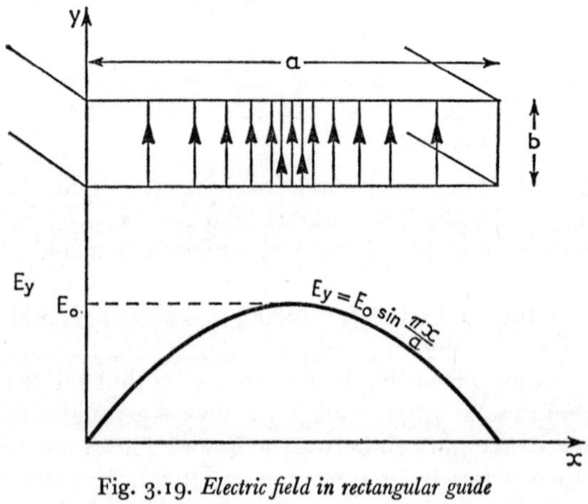

Fig. 3.19. *Electric field in rectangular guide*

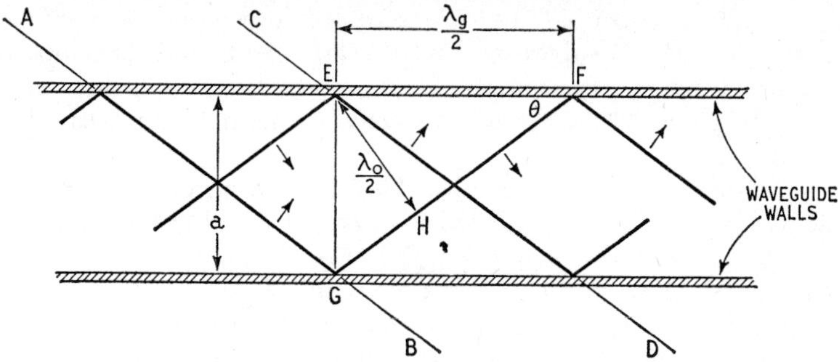

Fig. 3.20. *Waveguide propagation by reflection from the walls*

π radians apart in phase. Then since EF is a half-wavelength measured in the direction of the waveguide

$$\tan\theta = \frac{2a}{\lambda_g}$$

and since *EH* is one half the free space wavelength

$$\sin\theta = \frac{EH}{EF} = \frac{\lambda_o}{\lambda_g}$$

Thus

$$\cos\theta = \frac{\sin\theta}{\tan\theta} = \frac{\lambda_o}{2a}$$

and using $\cos^2\theta + \sin^2\theta = 1$ we get

$$\left(\frac{\lambda_o}{2a}\right)^2 + \left(\frac{\lambda_o}{\lambda_g}\right)^2 = 1 \qquad (3.14)$$

Hence

$$\lambda_g = \frac{\lambda_0}{\left[1 - \left(\frac{\lambda_0}{2a}\right)^2\right]^{\frac{1}{2}}} \qquad (3.15)$$

It is apparent that at the cut-off wavelength $\lambda_o = 2a$ and λ_g is infinite, and so the phase velocity is infinite, as is shown by the $\omega - \beta$ diagram. At higher frequencies the guide wavelength decreases and approaches the free-space wavelength, while the group and phase velocities both approach the velocity of light.

3.14 Impedance Transformation in Waveguides

As is the case with coaxial lines, short-circuited lengths of rectangular waveguide are frequently used as resonant cavities, either for the measurement of frequency or to provide regions of high field strength for use in valves, and all the arguments and descriptions of standing wave ratio, impedance transformation, etc, which were given previously for lines, apply directly to rectangular waveguides, but with the proviso that the guide wavelength λ_g is the criterion for determining resonant lengths. This has the effect that whereas a short-circuited coaxial line will resonate if excited at the open end at the frequencies at which its length is $\lambda/4$, $3\lambda/4$, $5\lambda/4$, etc, and these frequencies will be in the ratio 1 : 3 : 5, this simple ratio would not be obtained with a rectangular waveguide, or indeed with any other dispersive circuit. It is also worth remarking that it is only by using the impedance-transforming properties of a terminated guide that it is ever possible to obtain a high impedance. If a waveguide is 'open circuited', i.e. left with an open end, this end does not look at all like a high impedance, but rather like a reasonably matched microwave aerial, and power is radiated quite well from the open end into free space. If for any reason it is desired to produce a high impedance at any point within the guide, this can only be done by placing a very low impedance, e.g. a short circuit, a quarter of a guide wavelength away.

The electric field variation in a waveguide has already been described. The currents which flow in the waveguide walls also produce magnetic fields inside the waveguide, and the complete pattern is as shown in Fig. 3.21.

3.15 Fast-wave and Slow-wave Circuits

Coaxial lines and rectangular waveguides are examples of *fast-wave* circuits. In both of them the phase velocities are of the order of the velocity of light, and when they are used in valves other than merely for carrying power from one point to another it is usually in the form of short-circuited resonant sections for valves using localised interaction, described in Chapter 7. In many microwave valves it is necessary to produce fields which are moving at much lower velocities. This is done by means of *slow-wave* circuits, in which the phase velocity over at least a range of frequencies can be suitably reduced. In general these circuits are described in connection with the valves in which they are used, but as an introduction for the sake of completeness a basic example of a *slow-wave periodic circuit* will be included here.

This circuit, which is shown in Fig. 3.22, has a behaviour basically very much like that of the set of coupled resonant circuits shown

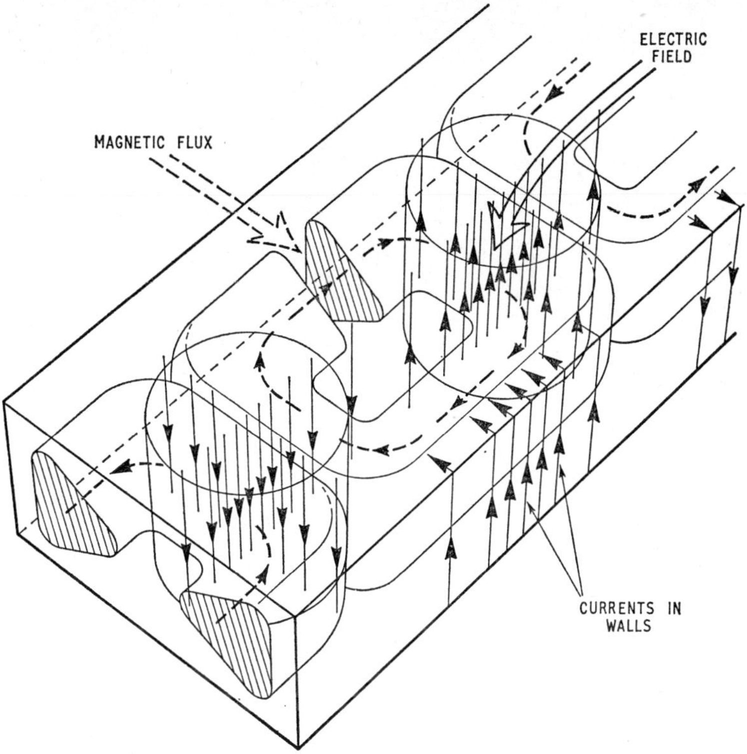

Fig. 3.21. *Pattern of fields in rectangular guide*

below it. These circuits comprise a band-pass filter, having a low attenuation for a range of frequencies around the resonant frequency, and a high attenuation for all frequencies outside this band. As seen earlier, a single cavity behaves as a resonant circuit. If it is coupled to an adjacent cavity by a small hole in its end plate, the electric field spreads out slightly through this hole, and effectively provides a capacitive coupling to the next cavity. The propagation characteristics of such a periodic structure can be measured by observing the frequencies and field patterns of the resonances of a fixed length, short circuited at its end. For example, if a length made up of ten cavities is taken, this is found to have ten resonant frequencies across its pass-band, starting at the low-frequency end where all the cavities are resonating in the same phase, and giving a resonance whenever the total phase change across the ten cavities is an exact multiple of π, i.e. whenever the circuit is an exact number of half-wavelengths

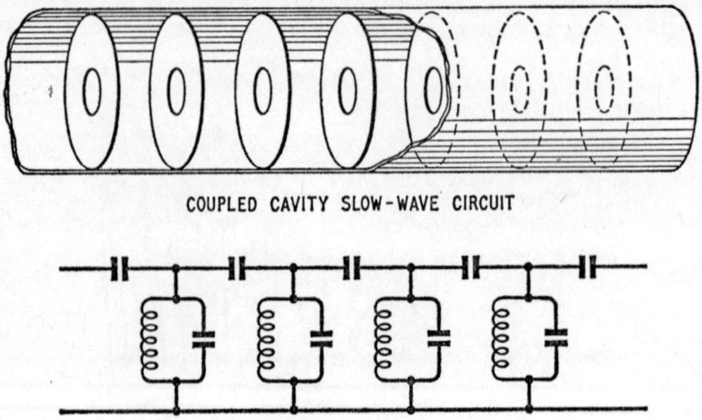

Fig. 3.22.

long. If the angular frequency ω is then plotted against the phase change per cavity, the $\omega - \beta$ diagram of this circuit is found to look as shown in Fig. 3.23.

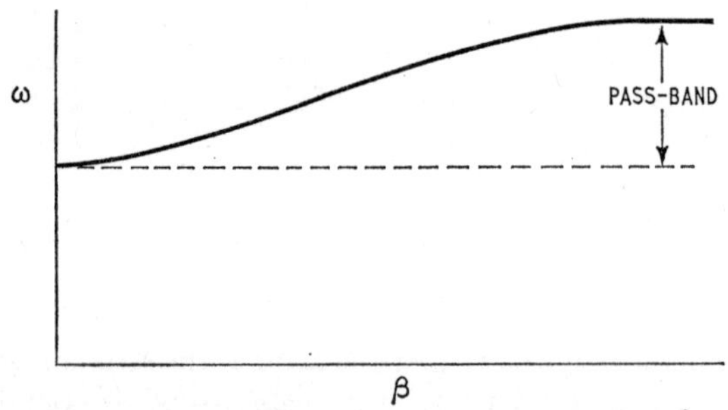

Fig. 3.23. ω-β *diagram of coupled cavity circuit*

From what has been said earlier it will be seen that the group velocity is zero at both ends of the pass-band, while the phase velocity starts at infinity at the low-frequency end, and gradually decreases across the pass-band, while both the phase and group velocities may approach the same value and remain roughly constant over the middle part of the pass-band.

BIBLIOGRAPHY

JACKSON, *Sir* WILLIS, *High Frequency Transmission Lines*, Methuen (1951).
RAMO, S., and WHINNERY, J. R., *Fields and Waves in Modern Radio*, Wiley (1953).

Chapter 4

LOW AND MEDIUM FREQUENCY VALVES

4.1 The Diode

Most of the microwave valves dealt with in later chapters only work over a limited frequency range. It is, however, a feature of diodes and of valves with grids that, although they can be designed to operate at microwave frequencies, these same valves will also work down to the lowest audio frequencies. In fact, their main use is at low and medium frequencies. The present chapter therefore deals with this use, in order to lay the foundation for extension to microwave use.

The simplest form of thermionic valve has only two electrodes, and hence is called a diode. One of these electrodes, called the cathode, can be heated to such a temperature that it emits electrons, and the other electrode, called the anode, will collect these electrons if its potential is made positive with respect to that of the cathode. If the anode potential is negative, it will repel the electrons and no current will flow.

The emission of electrons from the surface of a heated cathode may be likened to the liberation of molecules of steam from the surface of boiling water. In both the valve and the kettle the exciting force is heat. At ordinary room temperatures the number of electrons emitted by a cathode is so small that it may be regarded as zero; in the same circumstances evaporation from the surface of water takes place very slowly. There is continual activity amongst the molecules in a kettleful of cold water and amongst the electrons in the material of a cold cathode, but in both instances the surface forms a barrier which cannot be surmounted without the expenditure of more energy than is possessed by the great majority of the molecules or the electrons in question.

4.2 Thermionic Emission—The Potential Barrier

The effect of the heating is in each case to endow the particles with greater energy. When the molecules of water in a kettle or the free electrons in a cathode have acquired sufficient energy from the source of heat they are able to pass the surface barrier, 'paying for' their escape by the expenditure of some of this energy. The higher the temperature, the greater the emission. The barrier at the surface of a metal is an electrical one, due partly to the orientation of the atoms there, and partly to the local excess positive charge left when the electron is removed. This potential barrier varies considerably from metal to metal. It clearly must be as small as possible to obtain good electron emission. It is not, however, the only criterion to be considered, since for most metals the temperature required to give appreciable emission is either above the melting-point or is such that excessive evaporation takes place.

4.3 Types of Cathode

For these reasons the only pure metal which has had any real use as a thermionic emitter is tungsten (melting-point, 3,600°C). When this metal is run at 2,250°C (which is approximately the temperature of the filament in an electric light bulb) the electron emission is about 0·3 A/cm^2. Tungsten, in the form of directly heated filaments, was the material used for the cathodes of most large transmitting valves up to about 1945.

The potential barrier at the surface of a metal may be lowered by producing a thin film—actually about one molecule thick—of another suitable metal on the surface. For tungsten a film of thorium is used, the improvement being such that the running temperature can be reduced by 500°C for the same emission. This gives a reduction in filament heating power of about 65 per cent, a very real saving for the large valves concerned. In practice the surface film is produced by introducing a small amount of thorium oxide into the tungsten during manufacture and carburising the surface of the filament during processing of the valve. During operation slow chemical reaction occurs between the carbon and the thorium oxide, producing thorium which diffuses to the metal surface and continuously replaces that which is lost due to evaporation and other causes. The emission life of these valves is determined by the loss of the carbon, and will therefore be proportional to the ratio of volume to surface, i.e. to the diameter of the filament. For filaments 1 mm in diameter the life is many tens of thousands of hours. This type of emitter has considerable practical importance, and is used in most broadcast and medium-power transmitting valves.

The type of cathode used in all receiving-valve types found in radio and television receivers, and in some of the microwave valves discussed later, consists of a small nickel tube (Fig. 4.1), coated

Fig. 4.1. *Indirectly heated cathode*

on its outside surface with a mixture of the oxides of barium, strontium and calcium, and heated indirectly from the inside by an insulated tungsten filament. When run at 800°C mean currents up to 250 mA/cm^2 can be taken from such oxide-coated cathodes, over lives of the order of 10,000 hours. The total emission is, however, much higher and, for repetitive pulses, each a few microseconds long, currents up to 10 A/cm^2 can be taken. The active coating is obtained by spraying on mixed carbonates in a liquid suspension, and decomposing these carbonates to oxides by heating the cathode during evacuation and processing. After sealing off the valves, the emission usually has to be stabilized by taking current from the cathode. In spite of the simplicity of this processing, the chemistry of the oxide-coated cathode is very complicated. Ultimately the emission depends on barium at the surface of the coating, but the junction between the base metal and the coating, as well as the nature of the conduction in the coating, are also determining factors.

For some purposes cathodes which are mechanically and electrically more robust than the oxide-coated cathode are required, and some increase in heating power can be tolerated. One such form of cathode is given by mixing nickel powder with the oxide coating, or by sintering nickel powder on to the cathode tube, and filling the pores with oxide coating. These cathodes are run at about 900°C and will give good lives with mean emission currents up to 1 A/cm². Another form of cathode has a porous tungsten body, which is either soaked in a suitable barium compound, such as the aluminate, or is mounted close to a pellet of such a compound, so that, when heated, the volatilized products diffuse through the pores to the surface. These cathodes must be run considerably hotter than the oxide-coated cathode. At 1,100°C continuous currents of a few amperes per square centimetre can be taken, with a good life, but they do not have the capacity to give much higher pulse currents. The robustness of both these types of cathode is due to the continuous metallic conduction to the surface of the emitter. They have particular application to many microwave valve types.

4.4 Temperature and Space-Charge Limitation

Let us consider what happens to the electron emission when any of these cathodes forms one electrode of a planar diode, that is a diode in which the two electrodes are plane parallel surfaces. The velocity of the emitted electrons varies between zero and a high value, although there are many more slow ones than fast ones; at a cathode temperature T°K the average velocity, expressed as an equivalent energy, is $6{\cdot}77 \times 10^{-5}\,T$ eV, which for an oxide-coated cathode at 1000°K becomes 0·07 eV. With low anode voltages, the electrons form a cloud in front of the cathode, which is called the *space charge*. Being negatively charged, it repels newly emitted electrons, slowing them down, and turning the slowest ones back to the cathode, so that only part of the emitted current reaches the anode. The anode current is then said to be *space charge limited*. The conditions may be represented by curves as in Fig. 4.2, which shows the potential distribution across the diode. With the anode and cathode at the same potential, as in the full line, the negative space charge forms a potential minimum in front of the cathode such that slow electrons are accelerated back to the cathode, while fast electrons, which pass the potential minimum, are then accelerated to the anode. The current reaching the anode under these conditions is quite small. It can, in fact, be calculated that, for an oxide-coated cathode of area 1 cm² and emitting 1 A, the current reaching an anode at a distance of 1 mm will be only 0·45 mA, and the potential

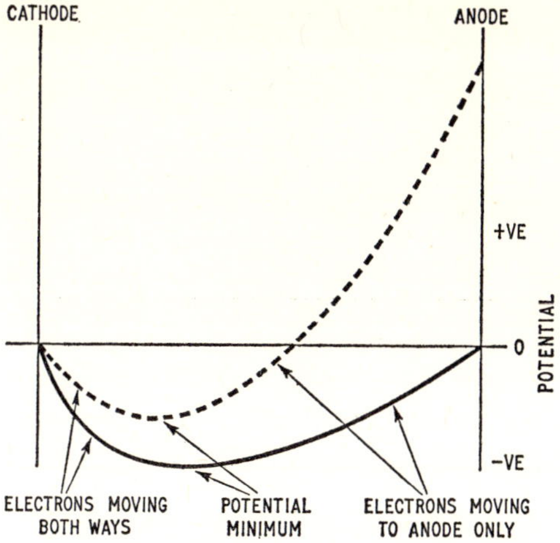

Fig. 4.2. *Potential distribution across diode*

minimum will have the value $-$ 0·7 V. If the spacing is reduced to 0·1 mm, the current will increase to 23 mA, and the potential minimum reduce to $-$0·34 V.

There is, as described in Chapter 2, a correspondence between the motion of electrons in electric fields and that of small bodies sliding on perfectly smooth slopes, i.e. moving in gravitational fields. To visualise this with the potential diagrams in this book, which are drawn with the usual electrical convention of positive potential upwards, the diagrams should be turned upside down. For example, Fig. 4.2 then shows a hill, which will act as a barrier, only letting over the fastest of the bodies shot from the cathode.

If the anode potential is made more positive, more current will flow to the anode, there will be less space charge, and the potential minimum will not be so deep, as indicated by the dotted line in Fig. 4.2. In terms of the analogy, the slope on the anode side of the hill is made steeper, and the hill is not so high.

With increasing anode potential all the electrons will ultimately be drawn to the anode. If the anode current is measured for many values of the anode voltage, using a circuit as shown in Fig. 4.3, the figures obtained may be plotted as in the curves of Fig. 4.4. The upper flat part of these curves, where most or all of the cathode emission travels to the anode, is shown dotted, because with most valves it cannot be measured with a simple circuit using a voltmeter

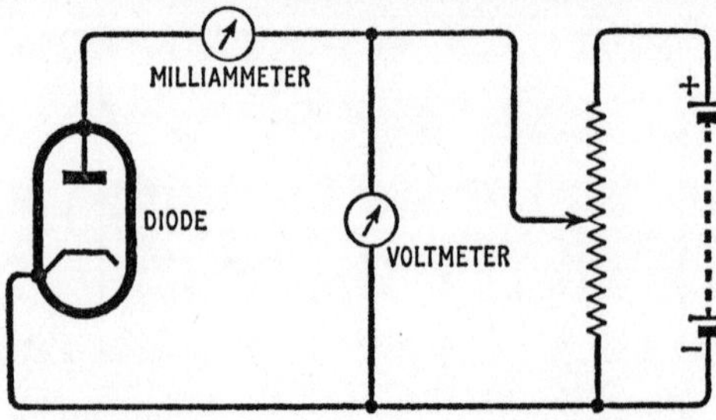

Fig. 4.3. *Circuit for measuring diode characteristics*

and ammeter, the power dissipated in the anode being too great. As shown, the current over this part of the curves is very dependent on cathode temperature, and is therefore said to be *temperature limited*, in contradistinction to that over the lower part which is space-charge limited. With an oxide-coated cathode, it only requires an increase of about 60°C in cathode temperature to double the emission. At the same time the evaporation of active material from the cathode will also increase sharply and therefore its life will be shortened. It can generally be taken that the manufacturers' heater rating is an optimum compromise between emission and life for most uses of valves.

Over its usable part, the current characteristic of the diode is

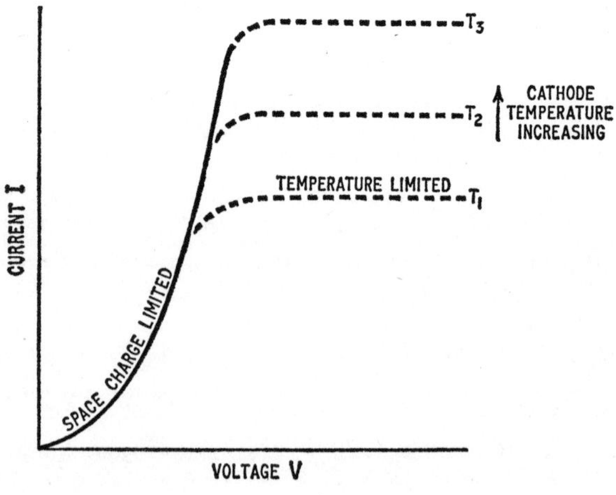

Fig. 4.4. *Diode characteristics*

curved upwards. In fact, over the central part of the range the relation of current density to applied voltage approximately follows a three-halves power law, given as an equation by

$$I = 2{\cdot}34 \times 10^{-6}.\ V^{3/2}d^{-2}\ \text{A/cm}^2 \qquad (4{\cdot}1)$$

when V is measured in volts and d is the electrode spacing in centimetres. Thus for a spacing of 1 mm and a voltage of 100 V a diode with an area of 1 cm² will pass a current of 234 mA.

4.5 Use of Diodes for Power Rectification

The uses of the diode arise from its property that it conducts electricity much more when the anode is made positive with respect to the cathode than when the anode is made negative. In fact, as we have seen, for negative potentials greater than about 0·5 V the current passed can usually be neglected, and the diode behaves as a true one-way valve. This property enables diodes to be used for changing alternating current into direct current, a process called rectification. Thus with the circuit of Fig. 4.5 the current through the load R can only flow in one direction, and a representation in

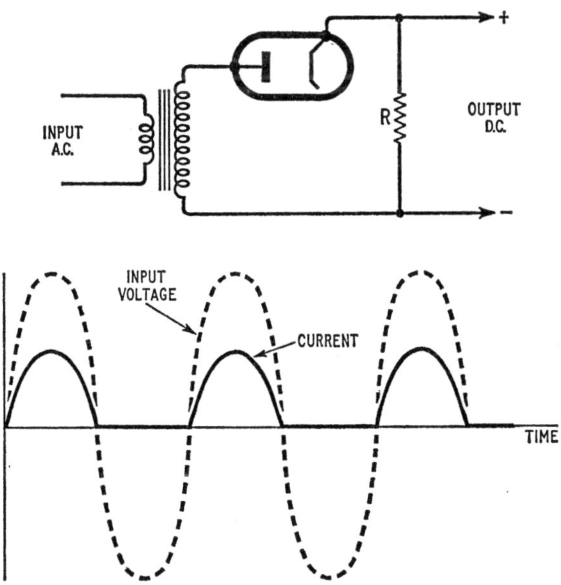

Fig. 4.5. *Operation of diode rectifier*

time of the voltage across R will consist of a series of humps as shown. To convert these unidirectional spurts of current into smooth d.c. the first step is to connect a large capacitor across R to act as a reservoir. After this capacitor has charged up during the first few cycles, its effect is to maintain the load current during the negative half cycles of the applied a.c. and to absorb the diode current during the positive half cycles, so that the load current now has the waveform of Fig. 4.6, i.e. d.c. with a small a.c. ripple of mains frequency

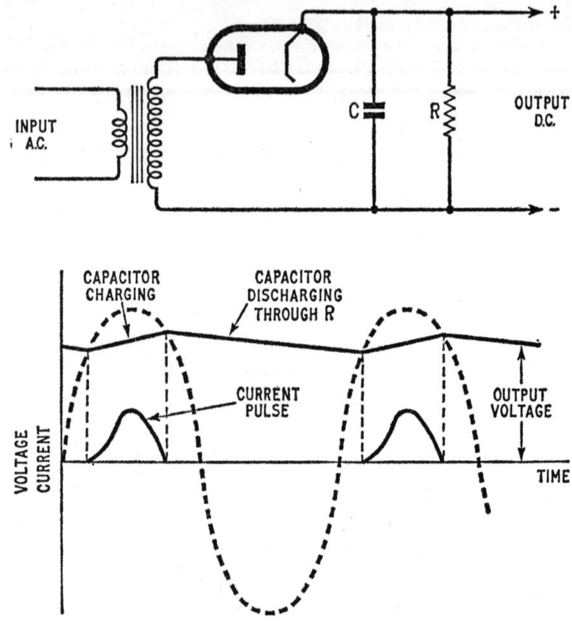

Fig. 4.6. *Rectifier with capacitive load*

superimposed. With an infinite capacitance, the current would be constant, but an approach to this condition is obtained more cheaply by using a further smoothing circuit consisting of inductance (in the form of an iron-cored choke) and capacitance as in Fig. 4.7. The inductor has high impedance to the ripple current and causes this to be bypassed through the low impedance capacitor and never reach the output.

There are two points which affect the diode which should be noted from Fig. 4.6. The first is that the diode cathode is main-

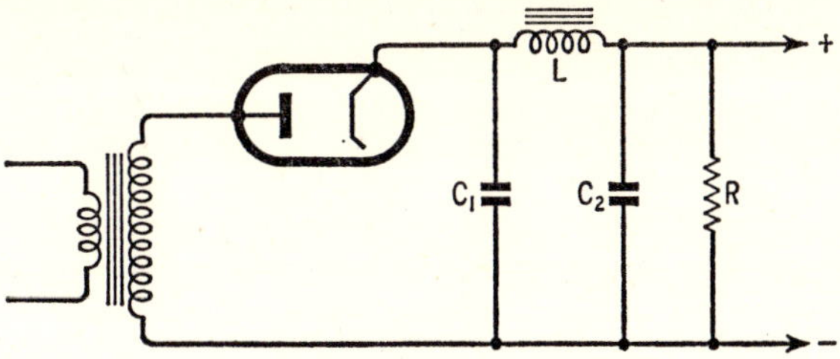

Fig. 4.7. *Rectifier with smoothing*

tained all the time at a high positive voltage, close to the peak positive value of the a.c. input voltage. Once a cycle, therefore, when the input voltage has its peak negative value, the total voltage across the diode is approximately twice the peak value of applied voltage. For example, with 240 V a.c. applied, the maximum voltage across the diode approaches $2 \times 240\sqrt{2}$ V = 680 V. The second point is that the diode only conducts when the instantaneous value of the applied a.c. voltage exceeds the capacitor voltage, which is for appreciably less than half the cycle, and therefore the peak current for a given mean current must be considerably higher than without the condenser. The ratio of peak to mean current in practice usually lies between 6 and 8.

This form of circuit, the so-called single-phase half-wave rectifier, is commonly used in radio and television receivers connected straight to the mains supply without an intermediate transformer. For higher output voltages a transformer must be used, and then it is better to use the biphase half-wave circuit with two diodes (usually in one bulb) as in Fig. 4.8 The advantage of this circuit is that conduction through the valve takes place twice per cycle and the ripple frequency is 100 c/s, so that for the same degree of smoothing the

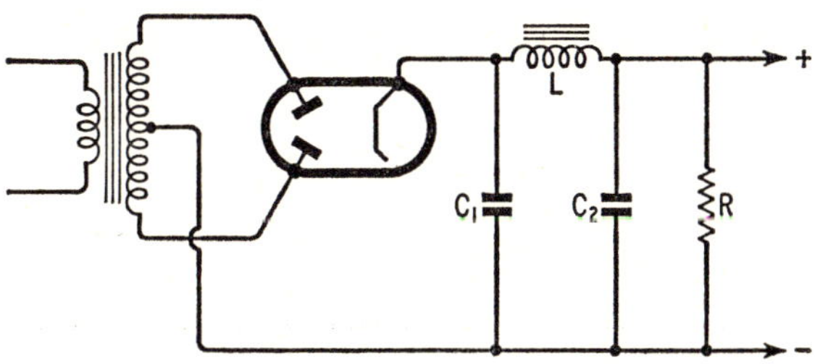

Fig. 4.8. *Biphase half-wave rectifier*

inductors and capacitors need have only half the values used in the single-phase circuit.

4.6 Signal Detection

The single-phase half-wave circuit is also used for the rectification of radio-frequency waves, the process then more usually being called detection, because it extracts or detects the information available in the modulated r.f. wave applied. To understand this we must divert for a moment to a consideration of how information (i.e. speech, music, etc) is conveyed on a r.f. current or voltage wave. This is done by altering or modulating some property of the r.f. wave in a manner corresponding to the waveform of the speech or other signal to be carried. In the present case we are concerned with amplitude modulation. For example, Fig. 4.9(a) shows a radio-frequency carrier whose amplitude is modulated by an audio-frequency wave. Such a modulated wave, if passed direct through a loudspeaker would give no audio output, as the movement cannot follow the rapid alternations in the r.f. current.

To detect the modulation, it must be fed to a circuit as in Fig. 4.10, which is similar to Fig. 4.6, but with the positions of the diode and the capacitor interchanged, so that the diode cathode can be

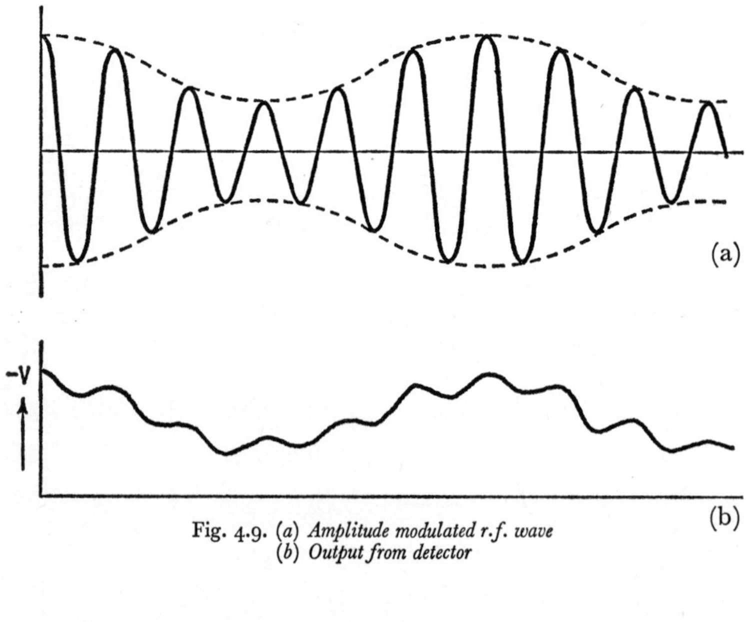

(b)

Fig. 4.9. (*a*) *Amplitude modulated r.f. wave*
(*b*) *Output from detector*

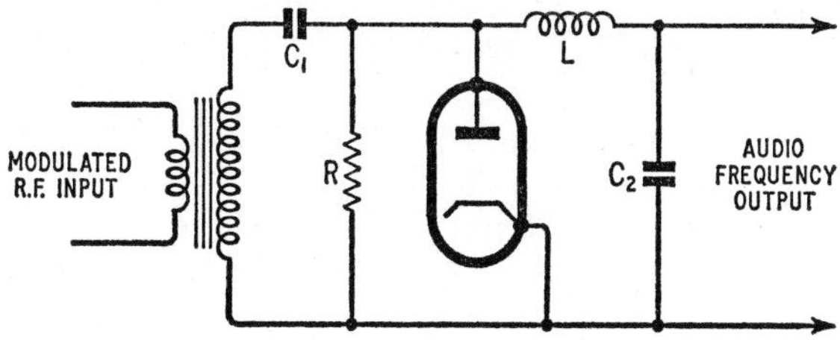

Fig. 4.10. *Detector circuit*

joined to the earthed end of the circuit, and thus keep r.f. out of the heater circuit. If the capacitor C_1 is of such a size that it is a low impedance at the radio frequency and a high impedance at the audio frequency, then the rectified voltage across the output capacitor will be an audio-frequency wave with a small r.f. ripple on it as in Fig. 4.9(b). Thus if R is 0·2 MΩ, then for a 1 Mc/s carrier modulated at 2,500 c/s a suitable value of C might be 100 pF, which has an impedance of 0·6 MΩ at 2,500 c/s and 1,500 Ω at 1 Mc/s. Further smoothing readily removes the remaining r.f. ripple. The circuit has thus detected or recovered the modulation information of the r.f. carrier.

4.7 Frequency Changing

A further use for diodes is as frequency changers or mixers in superheterodyne receivers. In these, not only is the r.f. signal voltage of frequency f_1 applied to the diode but also a much larger voltage of a frequency f_2, which may be higher or lower than f_1, and which is obtained from a local oscillator. Because of the non-linear characteristic of the diode, the resultant current will include not only the original frequencies and their harmonics, but also sum and difference frequencies, of which the latter is usually the one required for subsequent amplification. This can be seen from Fig. 4.11, which shows the waves of frequency f_1 and f_2 (with $f_1 < f_2$) and their sum. Visually it appears that the signal wave is just causing a variation at frequency $f_2 - f_1$ in the amplitude of a larger sine wave, which is applied to a rectifier. The rectified current will therefore have variations at this difference frequency, called the intermediate frequency (i.f.), and if a circuit, tuned to this frequency, is placed in series with the diode, the voltage developed across it will be the i.f. signal required for amplification. A block diagram for a diode mixer is shown in Fig. 4.12.

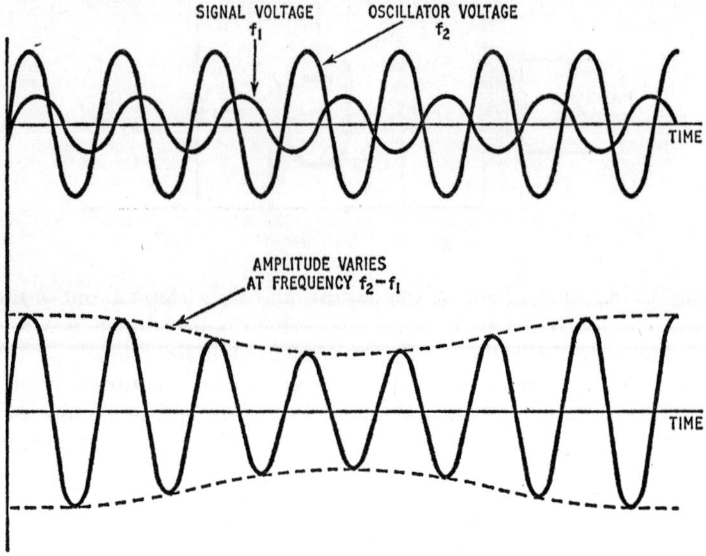

Fig. 4.11. *Voltages applied to frequency changer*

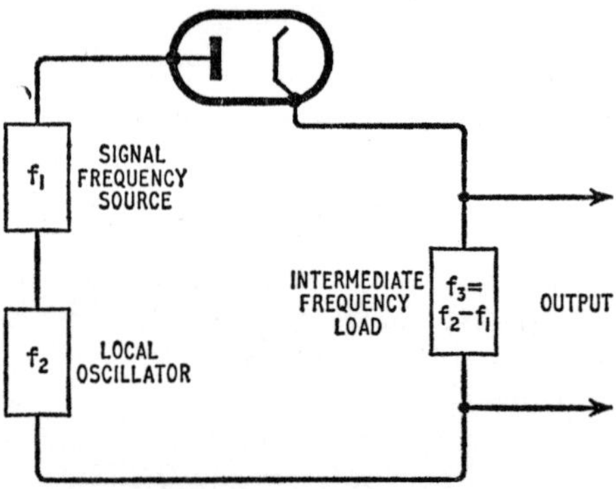

Fig. 4.12. *Diode frequency changer*

4.8 The Triode and its Characteristics

Useful though diodes are, they cannot be used for the generation and amplification of low- and medium-frequency power. For this, it is necessary to be able to control the current flowing to the anode, without altering the anode voltage. The commonest way of doing this is to introduce a third electrode, in the form of a grid of fine wires, into the path of the electrons. The resultant valve, which is known as a triode, is shown somewhat diagrammatically in Fig. 4.13. It consists of a cathode, shown indirectly heated, surrounded by a helix of wire usually wound on two supports, and this again surrounded by the anode. The winding pitch on the grid and its spacing from the cathode and anode are dependent on the type of valve involved.

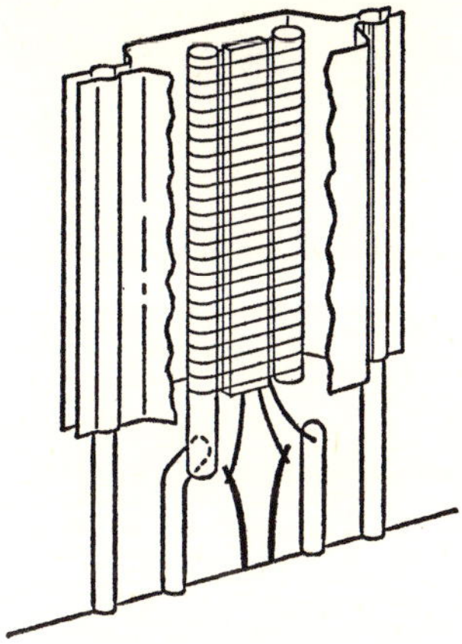

Fig. 4.13. *Structure of a triode*

The properties of the triode arise from the fact that a given potential on the grid is much more effective in drawing current from the cathode than the same potential on the anode. This would be expected intuitively, since the anode is farther away from the cathode, and also its effect has to penetrate through the grid. The valve behaves, in fact, as though there were an effective potential V_e in the plane of the grid given by the expression

$$V_e = p\left(V_g + \frac{V_a}{\mu}\right) \qquad (4.2)$$

where V_g and V_a are the grid and anode potentials, and μ and p are constants whose values depend on the grid dimensions and the interelectrode spacings. μ, which is called the amplification factor, is an important constant for a triode; for grid wires of diameter d cm, wound n to the cm, and spaced S_2 cm from the anode, an approximate formula for μ is

$$\mu = 53n^2dS_2 \qquad (4.3)$$

Thus, for example, a high value of amplification factor is obtained by using a close pitch on the grid and a large grid to anode spacing. The factor p usually lies between 0·5 and 1, having the lower figure for low values of μ.

We have seen that, over the central part of the characteristic of a diode, the anode current is proportional to the three-halves power of the anode potential. For the triode, the effective potential above must be used in the equation, so that the cathode current density will be given by

$$I = 2{\cdot}34 \times 10^{-6} \frac{p^{3/2}}{S_1^2} \left(V_g + \frac{V_a}{\mu}\right)^{3/2} \text{ A/cm}^2 \qquad (4{\cdot}4)$$

where S_1 is the distance (in cm) of the grid from the cathode. This current will be divided between the grid and the anode if they both have positive potentials. However, it will be seen that current can still flow when V_g is negative, if V_a/μ is greater than the numerical value of V_g, and in this case all the current will flow to the anode. It is in this negative grid condition that all the amplifying valves in radio and television sets work.

Consider a triode with an amplification factor of about 20 set up in a measurement circuit as in Fig. 4.14. With the grid voltage at

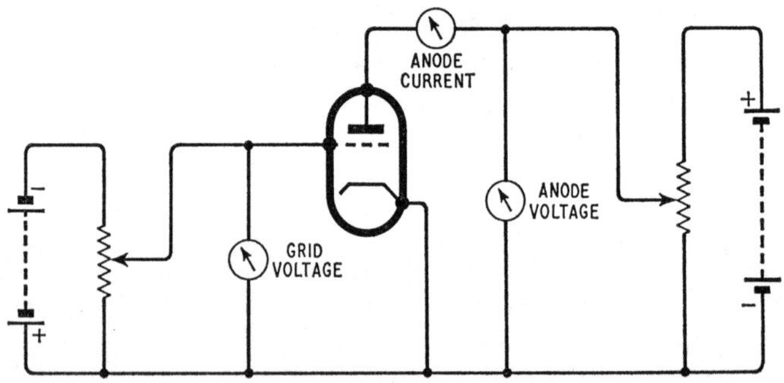

Fig. 4.14. *Triode measurement circuit*

zero, a curve can be drawn of anode current against anode voltage as at A in Fig. 4.15. This curve passes through the origin and, as would be expected from putting $V_g = 0$ in the equation above, follows approximately a three-halves power law, with an upward curvature. With -5 V on the grid, anode current will not flow until the anode voltage is quite high and the curve B has a similar shape to A, but is translated sideways by about 100 V ($=5\mu$). Similar

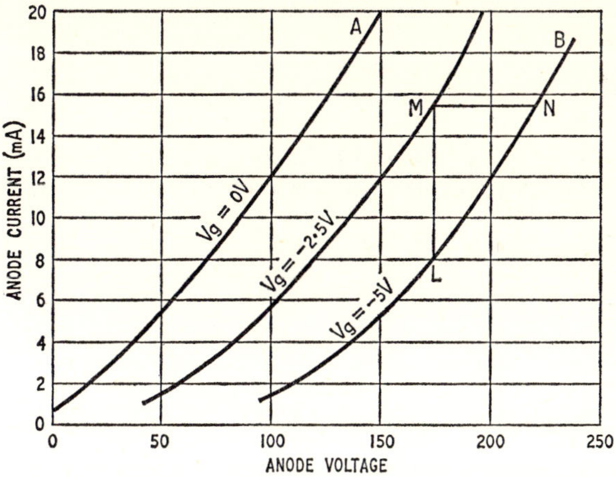

Fig. 4.15. *Triode characteristics*

curves for other grid voltages can be drawn, the whole forming a set of anode current-anode voltage characteristics. The same information could, of course, be plotted as characteristics of anode current against grid voltage, for various values of anode voltage.

4.9 Amplification Factor, Mutual Conductance and Anode Impedance

The three parameters which control the performance of a triode can be derived from Fig. 4.15. The first of these, the amplification factor μ, has been mentioned above. It is the ratio of the anode-voltage change which must be made in order to compensate for a small change in grid voltage and give the original anode current. If δV_a and δV_g denote the respective small changes in V_a and V_g,

$$\mu = -\left[\frac{\delta V_a}{\delta V_g}\right] I_a \text{ constant} \tag{4.5}$$

Thus, in going from M to N in Fig. 4.15 there is a change of +50 V in V_a and a change of −2.5 V in V_g. Thus

$$\mu = \frac{50}{2 \cdot 5} = 20$$

for this particular valve.

A second parameter is the mutual conductance, g_m, which is the ratio of the change δI_a in anode current to the small grid voltage δV_g

which produced the change, the anode voltage being maintained constant.

$$g_m = \left[\frac{\delta I_a}{\delta V_g}\right] V_a \text{ constant} \tag{4.6}$$

It will be seen that in going from L to M, a change in V_g of 2·5 V produces a change in current of 7·5 mA, i.e. the mutual conductance is 3·0 mA/V.

The third parameter is the anode resistance, r_a, which is the ratio of the change in anode voltage δV_a to the small change δI_a in anode current it produces, while the grid voltage remains constant, i.e.

$$r_a = \left[\frac{\delta V_a}{\delta I_a}\right] V_g \text{ constant} \tag{4.7}$$

In going from L to N, an anode current change of 7·5 mA is produced by a voltage change of 50 V, i.e. the anode resistance is 6,700 Ω.

These three parameters for a valve are not independent, since they involve the same changes of currents and voltages, the relationship between them being

$$\mu = g_m r_a \tag{4.8}$$

as can be checked from the numerical values quoted, using consistent units. It will be seen from Fig. 4.15 that because of the curved characteristics the mutual conductance falls off at low currents and the anode resistance correspondingly increases. Their product, however, remains roughly constant, as would be expected if the amplification factor depends only on the valve dimensions.

4.10 Triodes as Amplifiers—the Load Line

The use of triodes as a.c. amplifiers is illustrated by Fig. 4.16, which shows a typical curve of anode current plotted against grid voltage. A d.c. negative bias voltage is applied to the grid to bring the anode current to the operating point P. Superimposing an a.c. grid voltage then gives an a.c. variation in anode current as shown. At low frequencies this can be converted into a voltage variation by putting a load resistance R in series with the valve as represented in the circuit of Fig. 4.17. This circuit also shows how the negative bias may conveniently be obtained by the use of the resistance R_k in the cathode circuit, the shunting capacitance C_k being made of such a size as to smooth out voltage variations across R_k at the frequency being amplified (i.e. $\omega C_k R_k \gg 1$). For positive values of grid voltage,

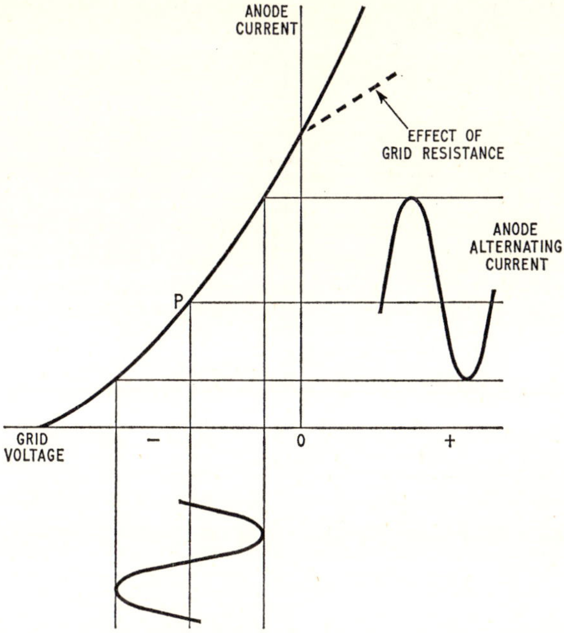

Fig. 4.16. *Principle of a.c. amplifiers*

when grid current flows, some voltage drop will occur across any series resistance present in the grid circuit, and the anode current will then rise more slowly, as along the dotted line in Fig. 4.16, which would cause distortion in the a.c. output. Unless, therefore, the resistance in the grid circuit is very low, the a.c. variations applied to the grid should be limited to the negative grid voltage part of the characteristic.

An increase in the anode current, produced by a positive change of applied grid voltage, will increase the voltage drop across the resistor *R*, and therefore reduce the voltage actually at the anode, which will tend to reduce the original current change. This effect can best be shown by drawing a 'load-line' for a resistor on the $I_a - V_a$ characteristics, as shown in Fig. 4.18. Suppose the series resistance has a value of 10,000 Ω, then for every 10 mA flowing the voltage drop across this resistor will be 100 V. With a 250 V supply, the resistance can, in effect, be represented by the load-line *OE*,

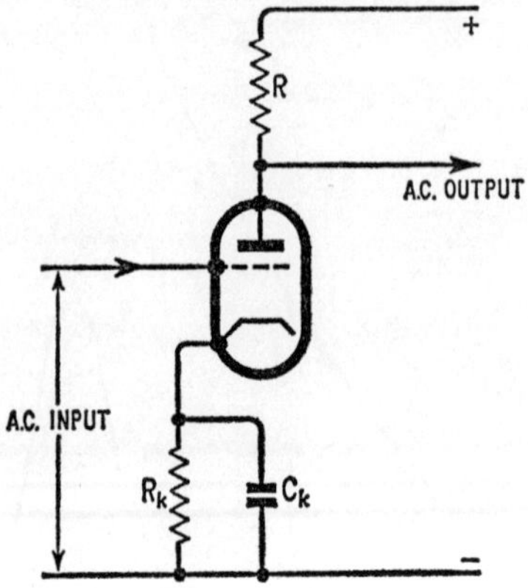

Fig. 4.17. *Low-frequency triode voltage amplifier*

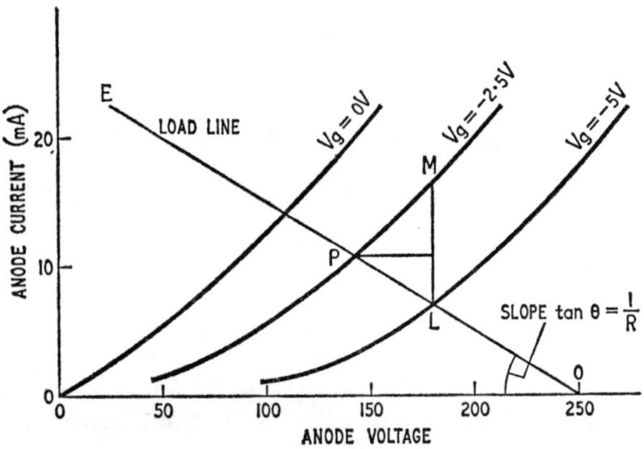

Fig. 4.18. *Load line on characteristics*

which is drawn backwards from the 250 V, 0 mA point with a slope of

$$\frac{10 \text{ mA}}{100 \text{ V}} = \frac{1}{10{,}000\ \Omega}$$

The operating point of the valve can only lie on this line, i.e. with a grid voltage change from −5 to −2·5 V, the anode current change is no longer from *L* to *M* but *L* to *P*, or 3 mA instead of the original 7·5 mA, with an anode voltage change of −30 V. The voltage gain under these conditions is then $30/2{\cdot}5 = 12$. In terms of the valve parameters, consideration of the geometry of Fig. 4.19 will show that the voltage gain is given by

$$A = \mu . \frac{R}{R + r_a} \tag{4·9}$$

$$= 20. \frac{10{,}000}{16{,}700} = 12$$

This expression for gain is exactly the same as would be obtained from the circuit of Fig. 4.19 with current being fed into the external load R from a circuit consisting of a voltage generator μv_g in series with a resistance r_a. This circuit is then said to be the equivalent anode circuit of the triode. It has wide application, one example being given in Chapter 5.

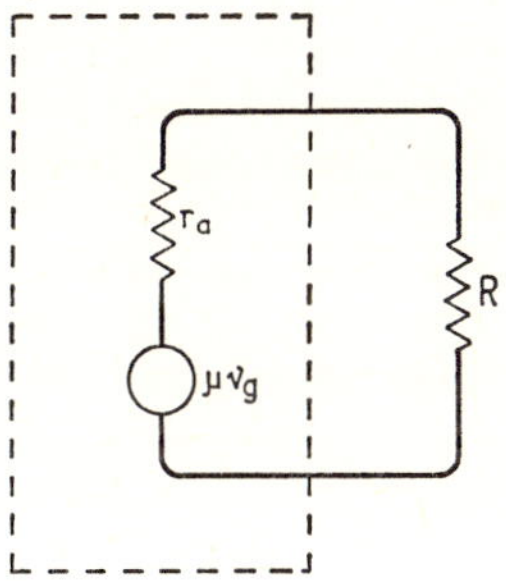

Fig. 4.19. *Equivalent circuit of a triode*

It is most important to note that the maximum current (i.e. most electrons) passes across the grid anode space when the anode voltage is a minimum, i.e. when the instantaneous a.c. field is in the direction to oppose the current flow. The electrons therefore have work done on them by the h.t. source, but give up energy to the a.c. field.

When a triode is used for power amplification rather than the voltage amplification above, the maximum voltage and current changes are wanted. In this case, the grid a.c. voltages need not be confined to the negative region, but may go positive. The characteristics of the above triode in this positive grid region are shown in Fig. 4.20. Apart from the greater power output obtainable, the only difference in operation compared with voltage amplification is that current is taken to the grid and has to be supplied by the drive.

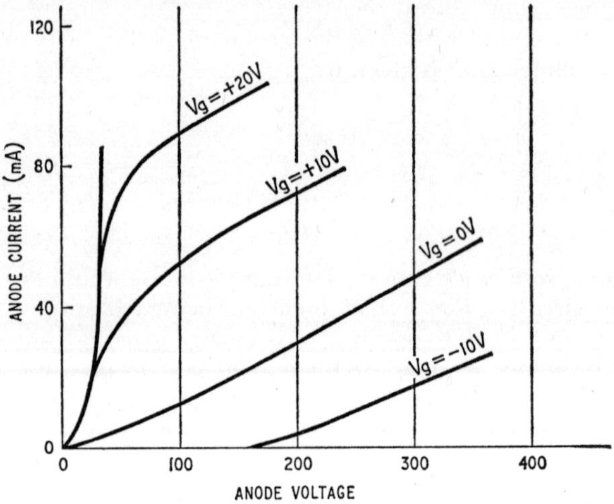

Fig. 4.20. *Triode positive grid characteristics*

4.11 Tetrodes

An improvement over the triode can be obtained if the functions of drawing current from the cathode and collecting it can be separated. This is done in tetrodes and pentodes by putting a second grid, called the screen grid, on the outside of the first (Fig. 4.21). With this second grid at a positive potential it draws the current from the cathode in exactly the same way as the triode anode, but a large proportion of the current passes through and will travel towards the anode. If the anode current is then measured for various values of anode voltage, keeping constant potentials on the two grids, it might be expected that with increasing potential the anode would rapidly collect all the current passing through the grid, to give

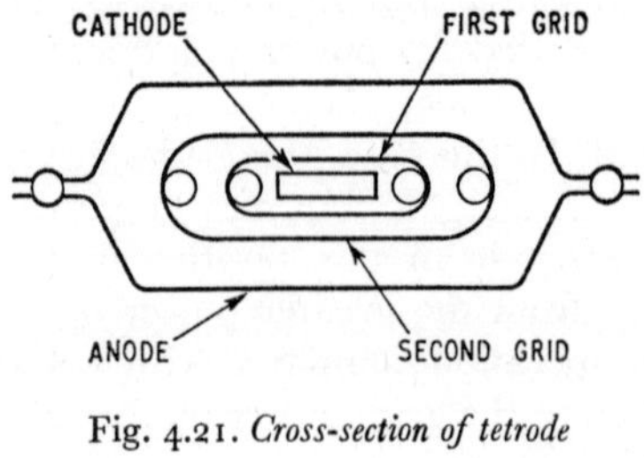

Fig. 4.21. *Cross-section of tetrode*

curves like those dotted in Fig. 4.22. In fact, curves like the full-line ones are obtained where the anode current rises at first, then falls and does not rise to its full value until the anode voltage is greater than that of the screen.

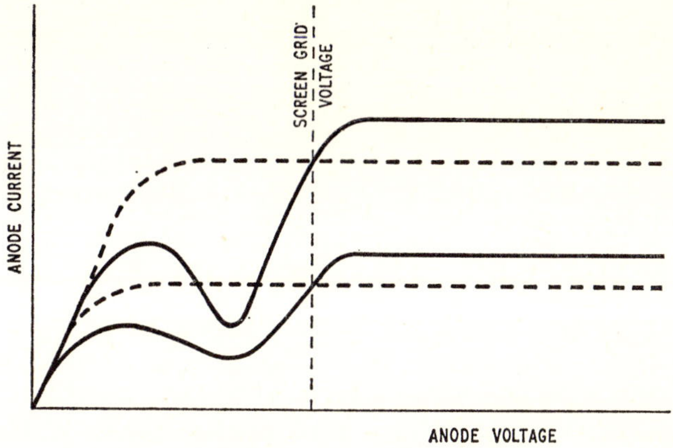

Fig. 4.22. *Characteristics of a screen-grid valve*

This form of characteristic arises from the important phenomenon known as secondary emission. When bombarded by electrons, all metals (and indeed, insulators as well, although we shall not be concerned with them) emit secondary electrons. A small fraction of these have approximately the energy of the bombarding electrons, being just reflected at or near the metal surface. The majority, however, have quite low energy, corresponding to a few volts, so that it is quite common to have the state where, on the average, each bombarding electron gives rise to many secondaries. The curve of the ratio (δ) of secondary emission current to bombarding current, plotted against the energy of the bombarding electrons—expressed in volts—takes the form shown in Fig. 4.23, which has a maximum at a few hundred volts. Higher energies make the bombarding electrons penetrate farther into the metal, and although more secondary electrons are produced, they cannot all reach the surface to escape, so that the ratio δ begins to decrease. Suitable preparation of the metal surface, such as carburisation, can considerably reduce the secondary emission ratio, but will never stop it completely.

The explanation of Fig. 4.22 can now be seen. With the anode voltage below the screen voltage, the slow secondary electrons from

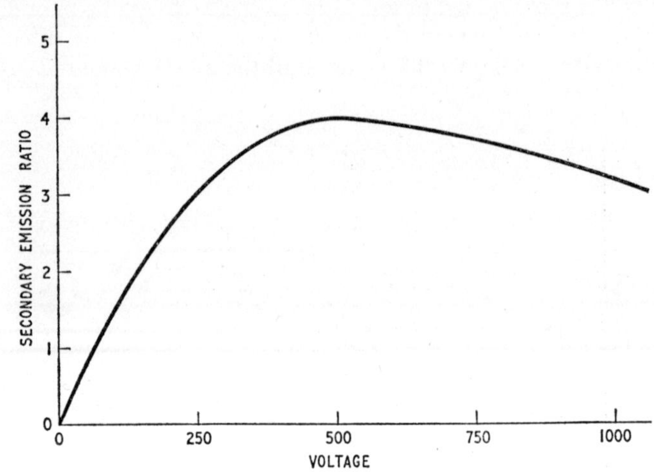

Fig. 4.23. *Variation of secondary emission with bombarding voltage*

the anode can travel to the screen, and since increasing anode voltage produces a bigger ratio of secondaries, the anode current first reaches a maximum and then decreases. It can sometimes even become negative, i.e. give a reversal of the normal direction of current flow in the anode lead. As the anode voltage approaches the screen voltage the anode current increases again, as space charge impedes the flow of the secondaries. With the anode voltage above that of the screen, the position is reversed, and the current reaching the anode is augmented by secondaries emitted by the screen.

4.12 Pentodes

These peculiarities in the characteristics of tetrodes, which may make them unsatisfactory to use, can be removed if the potential at a surface intermediate between the screen and the anode can be lowered to such a level that the secondary electrons produced at either electrode cannot escape from the vicinity where they are produced. This is carried out in practice by introducing a further electrode between the screen and anode, and operating this electrode at cathode potential. The electrode, shown in Fig. 4.24, can be either a complete grid, as in (a), wound with a very open pitch, or a pair of side plates as in (b). In either case there is very little effect on the flow of the main stream of electrons, but the effective

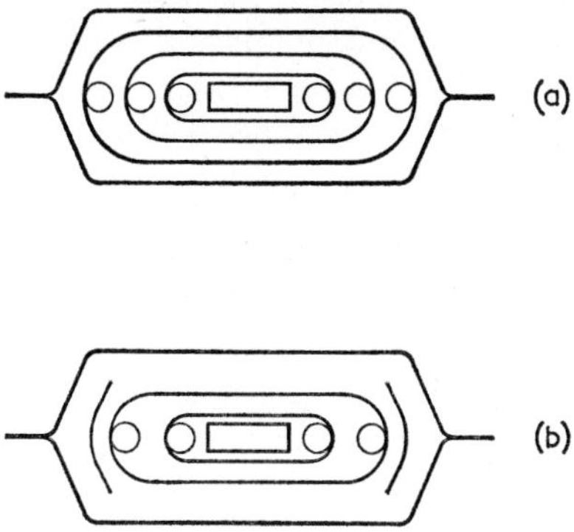

Fig. 4.24. *Alternative cross-sections of pentodes*

potential in the plane of the new grid, called the suppressor grid, is low enough to suppress the flow of the slow secondary electrons. The resultant characteristics of the five-electrode valve or pentode, as it is called, are shown in Fig. 4.25 for a fixed screen potential and various values of control grid potential.

The working part of these characteristics, as indicated by the load line, is where the curves are flat, with little inclination to the voltage axis. From the previous definitions, it will be seen that both

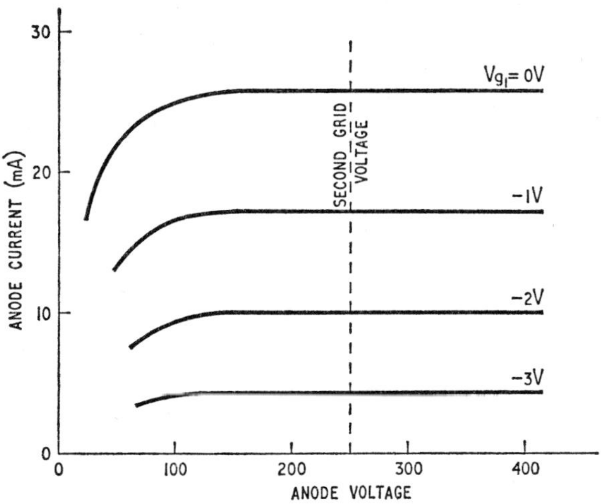

Fig. 4.25. *Characteristics of an r.f. pentode*

μ and r_a have very high values for such characteristics, and mean little to the operation of the valve. Their ratio, g_m, however, has a value similar to that obtained with triodes. With a large value of r_a, the voltage gain of a simple amplifier will be

$$A \simeq \mu \frac{R}{r_a} = g_m R \tag{4.10}$$

showing that the mutual conductance is the most important parameter for determining the voltage gain of a pentode amplifier, and that this gain can be much greater than with a triode. Also, as the $V_g = 0$ characteristic is usable down to a lower anode voltage with the pentode, a greater output voltage can be obtained.

The mutual conductance of common television-receiver type r.f. pentodes is about 10 mA/V at an anode current of 10 mA, while for similar-sized valves made for special purposes it can be as high as 45 mA/V at an anode current of 40 mA. To obtain such characteristics requires the use of grid wires smaller than 0·01 mm diameter, spaced at a distance of less than 0·1 mm from the cathode oxide coating.

4.13 Double Control Valves

The rising part of the characteristics of Fig. 4.25, where V_a is small, has similarities to the $V_g = 0$ curve of a triode. In fact, the anode and suppressor grid can be regarded as a triode operating on the current passing through the screen grid. This similarity becomes even stronger if the suppressor is wound with a closer pitch to give greater control of the anode current. However, as in the case of the triode, the amplification would be limited, and to use this outer part of the valve satisfactorily it is necessary to put in a further screen grid and suppressor grid between the third grid and anode. The resulting valve, a heptode (meaning seven electrodes) is represented diagrammatically by Fig. 4.26. The first and third grids are used as control grids, the second and fourth grids are called screen grids and are connected together to a positive voltage, while the fifth grid is an open-pitch suppressor grid, and is connected to the cathode. The d.c. characteristics of anode current against each grid voltage are shown in Fig. 4.27. In each case the different curves are for fixed values of the other grid voltage.

The major use of double control valves, of which the heptode is a typical, although not the only, type, is as frequency changers, with the signal and oscillator voltages connected respectively to the two

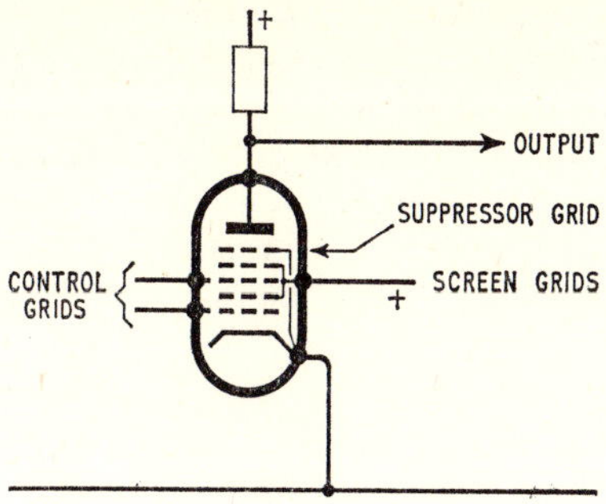

Fig. 4.26. *Diagram of heptode connections*

control grids. Examination of Fig. 4.27 will show that the application of simultaneous small voltage changes to both control grids will give a total change in anode current which is the sum of two parts which are respectively proportional to the voltage changes, together with a third part proportional to their product. As in the case of the diode (p. 58), this product term gives rise to sum and difference frequencies when the inputs are alternating voltages.

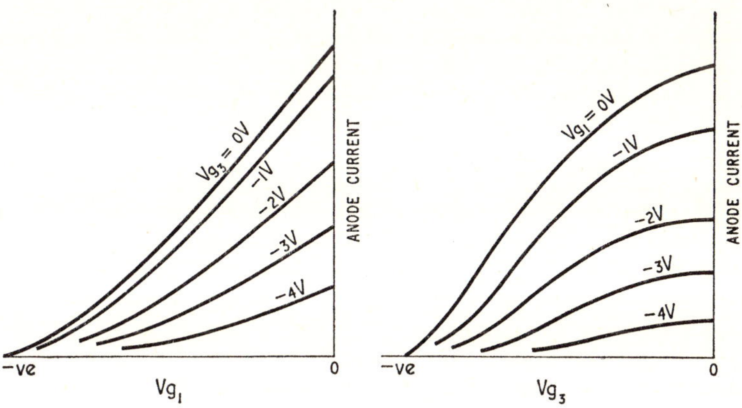

Fig. 4.27. *Heptode characteristics*

4.14 Use of Tuned Circuits

So far, in considering the amplification of valves, we have assumed that the load in the anode circuit is a resistance. When used for amplification at medium and higher frequencies—say 1 Mc/s upwards—it is usually required that the amplification should only occur over a particular range of frequencies. For this reason, and others dealt with later, the load then takes the form of a parallel tuned circuit, the properties of which have been treated in Chapter 3. When, as is usual, a pentode is used, the amplification at any frequency will be given by the product of the mutual conductance and the impedance of the tuned circuit at that frequency

$$A = g_m \cdot \frac{1}{\frac{1}{R} + j\omega C + \frac{1}{j\omega L}} \tag{4.11}$$

This gain has, of course, its maximum value of $g_m R$ at resonance, when

$$\omega L = \frac{1}{\omega C} \tag{4.12}$$

There is one striking difference between the use of resistive and resonant loads for amplification. Using the former, the waveform of the voltage across the load will be a reasonably faithful replica of that of the current passing through the valve. The resonant load, however, behaves like an electrical flywheel, as long as the dissipation per cycle in the load is small compared with the peak energy stored in the tuned circuit (i.e. the Q is high). The anode current variations only 'top-up' the energy in the tuned circuit, making up for the losses. It is not then necessary for the current variations to be sinusoidal, although they usually are in the very low power r.f. amplifiers used in radio and television sets where efficiency, i.e. the ratio of r.f. output power to the d.c. power supplied, does not matter. In high-power r.f. amplifiers, such as are used in transmitting stations, high efficiency can be obtained by using a d.c. negative grid bias so high that it would completely prevent anode current from flowing. The superimposed a.c. grid voltage will then cause the anode current to flow in spurts (Fig. 4.28) when the anode voltage is a minimum and less than the mean d.c. supply voltage. The power dissipated at the anode will therefore be considerably less than the power supplied, which is equal to the anode current times the mean d.c. voltage. The difference appears as r.f. power in the load. With suitable conditions, it can be arranged that more than 80 per cent of the supply power appears as r.f. power, and less than 20 per cent as dissipation in the valve anode.

This is an excellent example of the fundamental principle of operation of a.c. amplifiers, viz. that the electrons flow against the

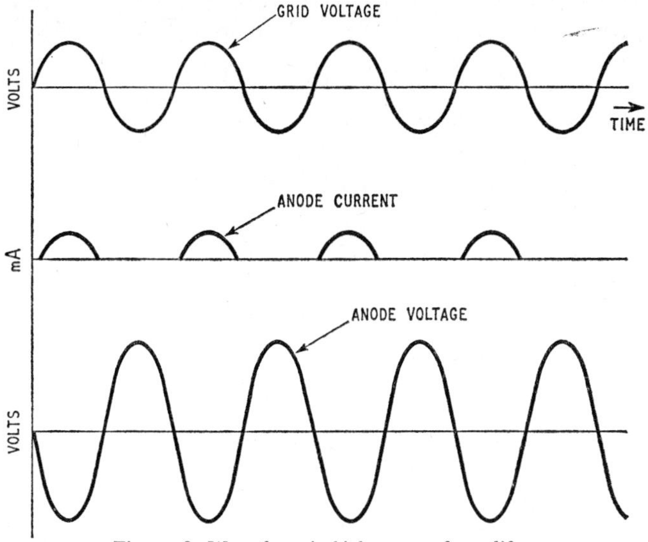

Fig. 4.28. *Wave-forms in high-power r.f. amplifier*

alternating field and are predominantly present in one phase only. This principle applies also to the types of microwave amplifier we shall meet later.

4.15 Self-Oscillation

Any valve which amplifies may also be used as a self-oscillator, i.e. as a generator of oscillations. It only requires that a portion of the output power should be fed back to the input to give the correct magnitude and phase of a.c. voltage at the grid which, when amplified, will give the original output. Thus, in the simple circuit of Fig. 4.29, a small coil in the grid circuit is coupled by mutual inductance to the anode-tuned circuit, which is connected via a blocking capacitor between anode and cathode, a choke being used to keep the r.f. power out of the h.t. supply circuit. The a.c. voltage fed to the grid will be proportional to the voltage across the tuned circuit, i.e. to the anode voltage, $V_g = kV_a$ say, where k represents the coupling between the coils. But if the impedance of the anode load is Z, then

$$V_a = Zi_a = Zg_m V_g = Zg_m kV_a$$

or

$$g_m k Z = 1 \tag{4.13}$$

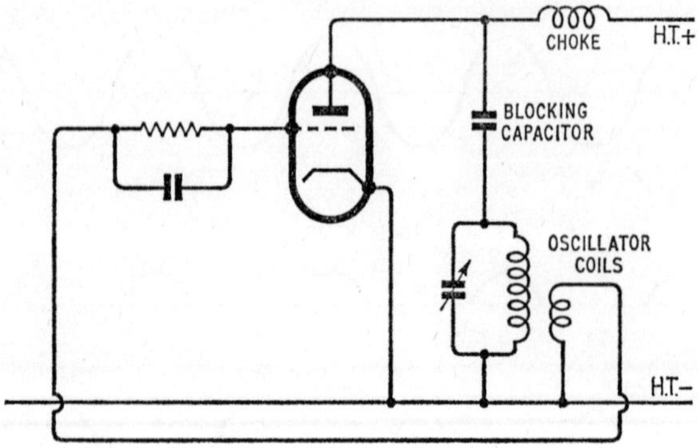

Fig. 4.29. *Simplified oscillator circuit*

which must be the condition for maintenance of oscillation. For a given circuit, i.e. k and Z fixed, the only variable is the mutual conductance g_m, and this must therefore be greater than a minimum value given by $1/kZ$. If g_m is greater than this, the oscillation amplitude will build up until some non-linearity in the grid circuit or the anode circuit or both causes an effective reduction in g_m to the required value. In the circuit of Fig. 4.29 it is the non-linear grid current characteristic which serves this role. Rectification takes place in the grid circuit, giving rise to a d.c. voltage across the grid load resistance R_g—called a grid leak—and the smoothing capacitor C_g. The resultant negative grid bias reduces the anode current, and therefore the mutual conductance, since the anode current characteristics are also non-linear.

This biasing action during oscillation ensures that the oscillation will always build up. At the moment of application of the h.t. voltage there will be no grid bias and, therefore, high current and high g_m. The minimum condition for oscillation is, therefore, more than fulfilled.

There must, of course, be some small oscillatory disturbance from which the oscillation can build up. This disturbance is provided by the noise voltages which we will discuss later in Chapter 13. Although these voltages are spread over a wide frequency band, the presence of the tuned circuit ensures that it is only at one particular frequency that the maximum voltage fed back to the grid is in the correct phase.

From consideration of this simple circuit the three basic properties which must be provided by any oscillator may be picked out. There must be

(*a*) sufficient amplification to overcome any circuit losses;
(*b*) a limiting action to stabilize voltage or current;
(*c*) a frequency-selective circuit to provide change of phase with frequency.

BIBLIOGRAPHY

BECK, A. H. W., *Thermionic Valves,* Cambridge University Press (1953).
ALDOUS, W. H., and APPLETON, *Sir* EDWARD, *Thermionic Vacuum Tubes,* Methuen (1958).
SPANGENBERG, K. R., *Vacuum Tubes,* McGraw Hill (1948).

Chapter 5

HIGH FREQUENCY AND MICROWAVE OPERATION OF DIODES, TRIODES AND PENTODES

5.1 Introduction

In the previous chapter no mention was made of any frequency limitations of conventional valves, apart from the use of tuned circuits to confine operation to a required band of frequencies. We are now concerned with what happens as the frequency of operation is increased. There are, in fact, three properties of valves which cause modification to the low-frequency characteristics, viz.

(*a*) interelectrode capacitance,
(*b*) lead inductance,
(*c*) transit time.

These come into play at successively higher frequencies, and will be dealt with in turn.

5.2 Effects of Capacitance

Because of the finite size of valve electrodes, the capacitance between them is appreciable. For instance, the typical small triode whose characteristics were quoted in Chapter 4 has capacitances

Grid to cathode	(C_{gk})	$= 1{\cdot}7$ pF
Grid to anode	(C_{ga})	$= 2{\cdot}0$ pF
Anode to cathode	(C_{ak})	$= 2{\cdot}6$ pF.

Now the impedance of a capacitance C at a frequency f is given by $1/2\pi fC$, so that at 1 Mc/s the capacitances above will have impedances of 95,000, 80,000 and 60,000 Ω respectively. At 50 Mc/s, however, these impedances drop to 1,900, 1,600 and 1,200 Ω respectively.

To understand the effects of the capacitances it is convenient to add them to the equivalent circuit of Fig. 4.19 to give that shown in Fig. 5.1. In this, impedances Z_g and Z_a, which may be resistances

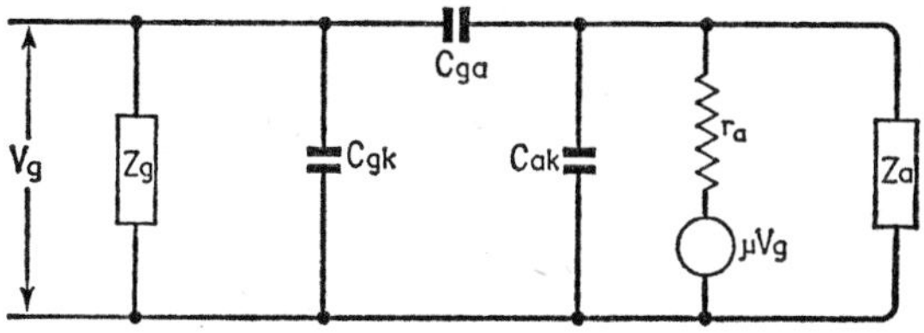

Fig. 5.1. *Equivalent circuit of triode at medium frequencies*

or tuned circuits, are shown shunted across the input and output respectively. At audio frequencies Z_a will usually be resistive and the impedance of the capacitances will be so high that they will have little effect. At television video frequencies (up to 6 Mc/s), however, the shunting effect of the anode-cathode capacitance makes it necessary to keep the anode load resistance, and therefore the gain, fairly low, in order to avoid a severe fall in gain with frequency. In this case inductance is often used in series with the load resistance, to form a low-Q tuned circuit with the valve capacitance, and thus extend the range of frequencies over which approximately constant load impedance can be obtained.

At radio frequencies up to, say, 30 Mc/s the valve input and output capacitances become part of the capacitances of their respective tuned circuits, and thus, for a given coil size, only affect the maximum frequency obtainable. There remains, however, the grid-anode capacitance which is, in effect, a coupling from the anode circuit to the grid circuit, and which can have much more serious effects than are at first apparent. It will be seen from Fig. 4.28 that at resonance the anode voltage waveform is in the opposite phase to the grid waveform. Thus with an alternating voltage V applied to the grid side of capacitance C_{ga}, and with a voltage gain A, an alternating voltage $-AV$ will appear on the anode side, and the current flowing will be as though a capacitance $(1+A)C_{ga}$ were connected across the input. With the triode quoted this would be much greater than the normal input capacitance even if the gain was only five.

5.2.1 Capacitive Feedback

There is another, and even more serious, effect of the grid anode capacitance, which occurs when the anode circuit is tuned off resonance and the anode voltage ceases to be in antiphase with the grid voltage. In this case, the current through the capacitance ceases to be in quadrature with the grid voltage. For frequencies below the anode circuit resonance this will produce a negative resistance effect at the input, with consequent regeneration, or even self-oscillation. Conversely, above resonance, the input circuit will be damped and the stage gain will be reduced.

While circuit means, known as neutralising, can be used to balance out the effects of this capacitive feedback over a small range of frequencies, a better way to obtain stable amplification is to reduce the value of the capacitance itself. This is what is achieved by the screen grid and pentode valves which were described in Chapter 4. In fact, the original purpose of the additional grids was for electrostatic screening between control grid and anode, although they also achieved the improved amplification already mentioned. The physical interposition of the screen grid and the suppressor grid, together with suitable arrangement of the connections, reduces the value of the anode-grid capacitance to about 0·01 pF, i.e. to about 1/200th part of its triode value. In consequence, at broadcast frequencies no special precautions have to be taken against feedback. At television frequencies the wider bandwidth required necessitates low gain per stage, as described earlier, in order to keep impedances low, and therefore feedback effects remain unimportant.

5.3 Effects of Inductance

It is at these television frequencies, e.g. above about 40 Mc/s, that the second of the properties of valves, viz. the effect of the inductance of the leads to the electrodes, begins to make itself felt. The self-inductance of a 1 cm length of the size of wire used for valve leads is approximately 0.01 μH and at 48 Mc/s this has the appreciable impedance of 3 Ω. If such a length of wire forms part of both the input and output circuits of a valve, then the voltage across the impedance due to the output current will appear also in the input circuit, giving feedback which affects the valve operation. This is what happens with the cathode lead of a high-frequency pentode, as shown in Fig. 5.2. The total current through the common inductance

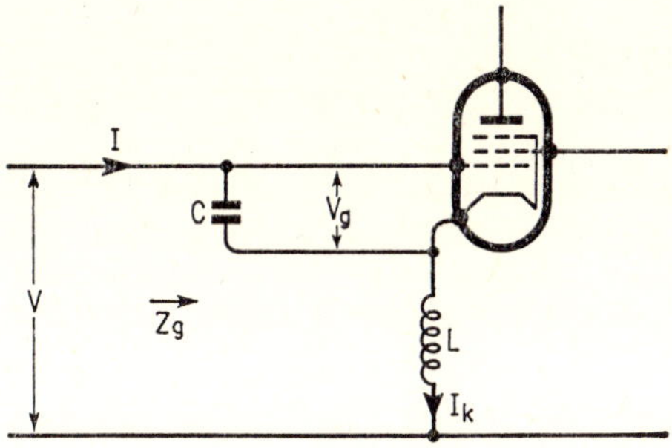

Fig. 5.2. *Circuit to show effect of cathode lead inductance L and grid-cathode capacitance C*

L is the sum of the valve current and the input capacitance current, i.e. for an angular frequency ω, and a voltage V_g appearing between grid and cathode,

$$I_k = g_m V_g + j\omega C V_g \qquad (5.1)$$

but

$$V = V_g + j\omega L I_k) \qquad (5.2)$$

and

$$V_g = \frac{I}{j\omega C} \qquad (5.3)$$

from which the input impedance Z_g is given by

$$Z_g = \frac{V}{I} = \frac{1}{j\omega C} + j\omega L + g_m \frac{L}{C} \qquad (5.4)$$

If this series circuit is transformed to an equivalent parallel circuit as in Fig. 5.3 the loss resistance shunted across the valve input by the effect of *L* is seen to be

$$R_g = \frac{1}{\omega^2 L C g_m} \qquad (5.5)$$

For example, at a frequency of 45 Mc/s, with $g_m = 0{\cdot}0075$ mho (i.e. 7·5 mA/V), $L = 0{\cdot}015$ μH, and $C = 10$ pF, R_g becomes 11,000 Ω which will exert considerable damping on a tuned circuit.

The effect of the cathode lead may be considerably reduced by using a push-pull system in one valve envelope, preferably with a

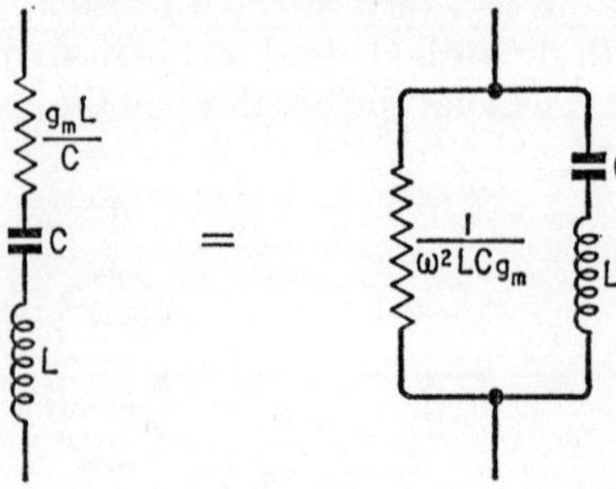

Fig. 5.3. *Equivalent series and parallel circuits representing input impedance*

common cathode, which removes the a.c. component of the anode current from the lead. Valves made in this manner will operate as amplifiers at frequencies up to many hundreds of megacycles per second.

An effect similar to the above will occur due to the screen lead in a pentode, but it will be smaller in the ratio of the screen current to the cathode current. The effect will also be in the opposite direction, producing negative resistance at the input, which, to some extent, offsets the damping effect of the cathode lead inductance.

5.3.1 Grounded-Grid Operation

Because of these effects it is natural to consider the use of a circuit in which the common lead (or leads) does not carry the anode current. Such is the grounded-grid or common-grid circuit, which is shown for a triode in Fig. 5.4. In fact, there is little advantage in

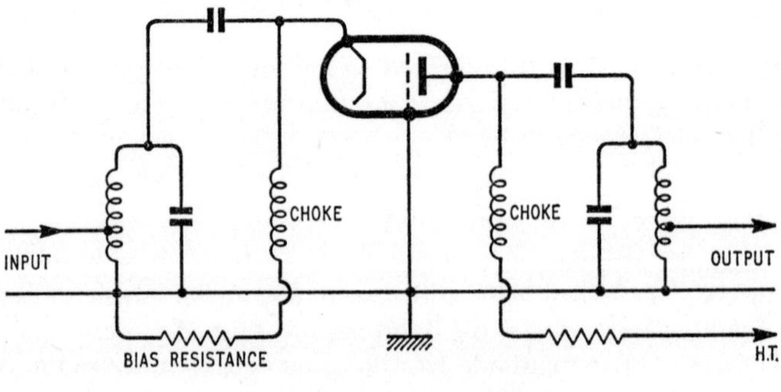

Fig. 5.4. *Grounded-grid amplifier circuit*

using a pentode in this form of circuit, since the triode grid now acts as the electrostatic screen between the circuits. At the frequencies where the grounded-grid connection is used the circuit impedances are low and the higher possible gain of a pentode, which always has a high impedance, cannot be achieved. The r.f. drive is applied to the cathode, so that the input and output circuits are in series and carry the same current. It is therefore a power-amplifying circuit and the power gain is equal to the voltage gain.

The grounded-grid connection is so successful that, by using multiple leads to the grid to reduce inductance, and suitable screens in the valve, miniature glass envelope triodes on the B9G base can be made which operate with gain and stability at frequencies up to 1,000 Mc/s. To carry the operating frequency range higher than this, special forms of construction are adopted, which minimize the inductance of the leads. Such constructions involve disc seals through the glass, and a typical valve is shown in cross-section in Fig. 5.5. With concentric line resonators as the tuned circuits such valves are suited for operation at frequencies up to 10 Gc/s.

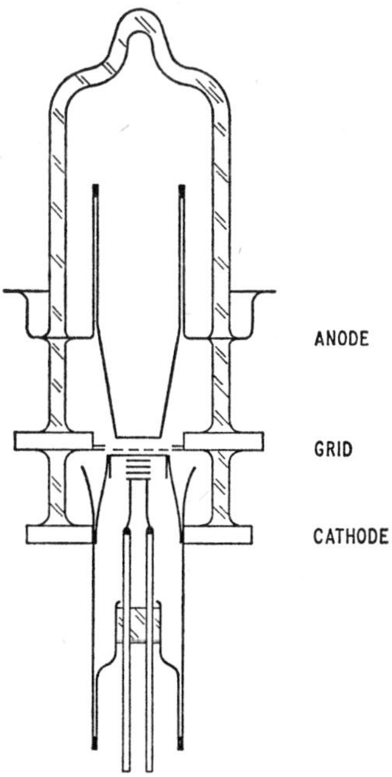

Fig. 5.5. *Cross-section of disc-seal triode*

5.4 Effects of Transit Time

At this very high frequency, and we are now very much in the microwave region, the third of the effects mentioned at the start of the chapter, viz. that of electron transit time, is the limiting phenomenon. The period of one r.f. cycle at 10 Gc/s is 10^{-10} sec, or 0·1 nanosec, in which time an electron can only move a very short distance away from the cathode. In consequence, the valve will not operate in the low-frequency manner described in Chapter 4.

5.4.1 The Temperature Limited Diode

To appreciate this, let us first consider the behaviour at high frequencies of a temperature-limited planar diode. By definition, the d.c. voltage on the anode produces a field sufficient to draw to the anode every electron emitted by the cathode, and at low frequencies an alternating voltage superimposed on the d.c. voltage would produce no corresponding alternating current. The transit time τ for such a diode can be derived from the expression given for the force on an electron in Chapter 1, viz. $F=-Ee$, the electric field E in this case being constant and equal to $-V/d$. Remembering Newton's laws of motion, we know

$$\text{Force} = \text{mass} \times \text{acceleration } (f)$$

that is

$$\frac{eV}{d} = mf \tag{5.6}$$

and distance travelled

$$d = \tfrac{1}{2} f \tau^2 \tag{5.7}$$

from which it follows that

$$\tau = d\sqrt{\frac{2m}{eV}} \tag{5.8}$$

Putting in values, say 10 V for a spacing of 3×10^{-4} m (i.e. 0·3 mm), then $\tau=3{\cdot}2\times10^{-10}$ s. The quantity which really affects the performance of the diode at high frequencies is the transit angle θ ($=\omega\tau$), which is the phase angle through which the signal changes during the transit. It is usually the case that valve properties are not appreciably affected if θ is less than one-tenth of a cycle, i.e. less than $\pi/5$ rad. For the diode dimensions above the dividing line would occur at 300 Mc/s.

Suppose we have an alternating voltage, $V_{\sim} \cos \omega t$, at a frequency above 300 Mc/s, superimposed on the d.c. anode voltage, V_o. Then the electrons, which still leave the cathode all the time at the average rate, travel faster at the peak of the alternating field and slower at the minimum. Thus the electrons which build up in the space when the field is a minimum get swept to the anode as the field increases, to give a current peak in advance of the voltage peak. From simple a.c. theory this means that at high frequencies the effective diode capacitance is increased and a resistive component appears in parallel with it. If the direct current through the valve is I, then the alternating current i can be shown to be given by

$$\frac{i}{I} = 2\frac{V_{\sim}}{V_o}\frac{1}{\theta^2}[(2 - 2\cos\theta - \theta\sin\theta)\cos\omega t - (2\sin\theta - \theta - \theta\cos\theta)\sin\omega t] \qquad (5.9)$$

which, for small values of the transit angle, i.e. lower frequencies, becomes

$$\frac{i}{I} = \frac{V_{\sim}}{V_o}\left(\frac{\theta^2}{6}\cos\omega t - \frac{\theta}{3}\sin\omega t\right) \qquad (5.10)$$

In this the first term is the resistive component and the second the quadrature capacitive component.

It is important to realise that the current round the circuit due to the transit of an electron across the diode does not start when the electron reaches the anode, but is a continuous current during that transit. If the electron velocity is u at any moment, the current flowing in the external circuit is eu/d, which has zero value when the electron leaves the cathode, and has a maximum when the electron reaches the anode, the current waveform being as shown in Fig. 5.6. Because of this the time lag between current and voltage is not the actual transit time, but an appropriate fraction of this.

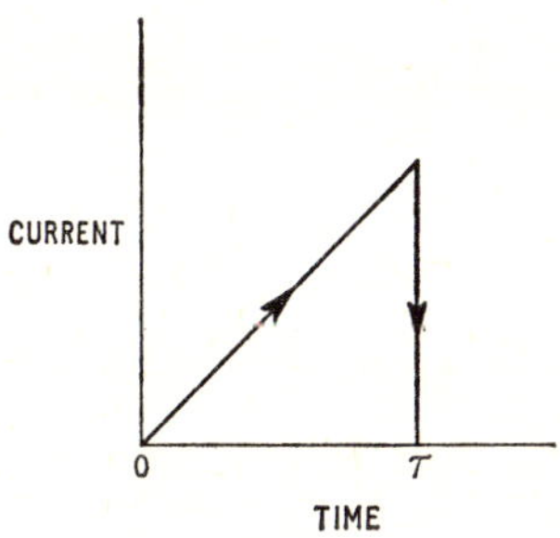

Fig. 5.6. *Induced current from single electron travelling across diode*

5.4.2 The Space-Charge Limited Diode

The case of space-charge limitation in the diode is the more usual one in practice, but the calculation of the effects of transit time is more difficult since, even for the ideal three-halves power law case, neither the emission nor the field strength are independent variables. What is done is to assume an impressed sinusoidal current and determine what alternating voltage is required at the anode to sustain this current. If the diode is represented by a capacitance C in parallel with a conductance G (where $G_0 = 3I/2V$ at low frequencies), it is found that the conductance decreases with increase of frequency and becomes zero when the transit angle equals 2π. This is because the variation in current taken from the cathode is balanced by variation in transit time. For transit angles between 2π and 3π the a.c. conductance goes negative, i.e. the instantaneous current decreases, while the voltage is increasing and vice versa.

A rather stranger thing is shown up by the analysis, viz, that the slight lag of current changes behind applied voltage changes has a distinct effect even down to zero frequency. In terms of circuit theory the space charge acts like an inductance. But for constant transit time the angular lag is $\omega\tau$ i.e. proportional to frequency, whereas a pure inductance gives a lag inversely proportional to frequency. A circuit element which gives a lag of the right form is a negative capacitance, and it turns out that its value is 0·4 C_0 where C_0 is the static capacitance of the diode. The low-frequency capacitance of a space-charge-limited diode is therefore 0·6 C_0. The variation with transit angle of C/C_0 and G/G_0 is shown in Fig. 5.7.

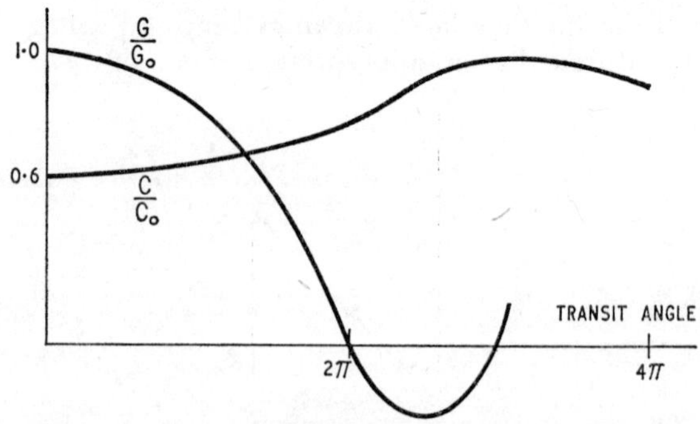

Fig. 5.7. *Variation of G/G_0 and C/C_0 with transit angle in space-charge limited diode*

It will be clear that to make use of diodes at microwave frequencies it is necessary to reduce the transit time to as low a value as possible. A diode made for use as a frequency changer for frequencies up to 10 Gc/s is shown in Fig. 5.8. This has an oxide-coated

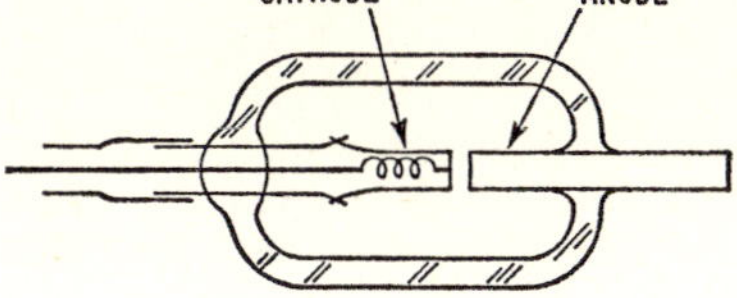

Fig. 5.8. *Cross-section of diode for use at 10* Gc/s

cathode with an area of 0·05 cm^2 spaced about 0·1 mm from the tungsten-rod anode. The capacitance is only 2 pF and with 1 V applied the anode current is 1 mA, at which level the transit time is only 5×10^{-10} sec.

5.4.3 The Triode

It might be expected that high-frequency effects similar to those in diodes would also occur between cathode and grid in negative-grid triodes. They do not show in quite the same way, however, since the resistive part of the grid-cathode impedance is infinite at low frequencies, i.e. no electron current flows to the grid. For small variations of potential on the grid, the electron trajectories at high frequencies will be negligibly different from those on d.c., and still no electrons will flow to the grid.

Transit time produces, in fact, two main effects. The first is the introduction of a phase angle into the mutual conductance. This effect is quite straightforward, in that the change in anode current, which is due to the effect of the grid voltage on the potential minimum, takes a finite time to reach the anode, and thus shows as a phase lag. This lag must be a function of the transit angles θ_1 and θ_2 across the cathode-grid and grid-anode spaces, and is given to the first order by $\frac{11}{30}\,\theta_1 + \frac{2}{3}\,\theta_2$. There will be, at the same time, a reduction in the magnitude of the mutual conductance, but this effect is small compared with the phase lag.

It may be noted in passing that a triode with a temperature-limited cathode, which, by definition, can have no mutual conductance at low frequencies, actually acquires this property at microwave frequencies.

The second effect results from the induced current in the grid circuit, and gives rise to low impedance and loss. It is akin to the change of impedance of the diode with frequency, but differs substantially in that no electrons actually reach the control grid. All pass through and travel with varying initial velocity across the space between grid and anode, which forms an effective second diode. The total current in the grid circuit therefore consists of the induced current from these two spaces. That, i_1, from the cathode-grid space must lag behind the applied a.c. field, while that, i_2, from the grid-anode space will be almost in phase opposition to i_1, but will lag slightly more. The resultant of these, as shown in the vector diagram of Fig. 5.9, is a small current i_g which leads the applied

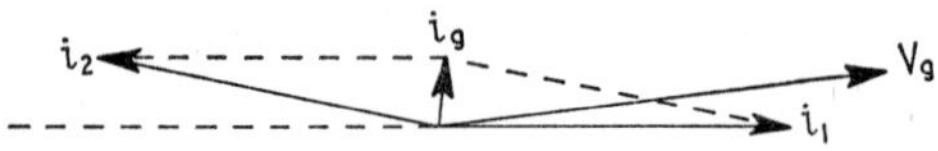

Fig. 5.9. *Vector diagram of current in grid circuit*

voltage. It can be resolved into capacitive and resistive components, the latter giving rise to the loss. A simplified mathematical formula for the input admittance y at high frequencies is

$$\frac{y}{g_m} = \frac{\theta_1^2}{20}\left[1 + \frac{44}{9}\frac{\theta_2}{\theta_1} + 5\left(\frac{\theta_2}{\theta_1}\right)^2\right] + j\frac{\theta_1}{6}\left(1 + 4\frac{\theta_2}{\theta_1}\right) \quad (5.11)$$

Thus for a typical small triode operating at a frequency of 150 Mc/s where θ_1 and θ_2 might be 1 rad and 0·3 rad respectively,

$$y/g_m = 0{\cdot}146 + 0{\cdot}37\,j$$

When the mutual conductance is 5 mA/V, the resistive part of the input admittance is then 1,400 Ω [$=1/0{\cdot}146 g_m$]. This leads to loss of gain, since a given drive power develops a lower voltage between grid and cathode, which produces even lower alternating current, and therefore lower power in the anode circuit.

By using grid-cathode clearances of the order of 0·04 mm small glass-based triodes are made which operate with a gain of about 10 dB up to 300 Mc/s. For higher-frequency operation it is necessary to use the grounded-grid circuit, where, although the input impedance is low, it is less dependent on frequency. With suitably placed multiple leads to give efficient grounding of the grid, and good internal screening, glass-based triodes can then be operated up to 1,000 Mc/s. The limit here is not, however, the transit time but the leads themselves, and disc-seal triodes with the same clearances will operate up to 3,000 Mc/s. With even closer clearances gain can be obtained above 10,000 Mc/s.

Operation of tetrodes and pentodes is also possible up to the lower microwave frequencies. No different principles are involved in this operation, and they will not be discussed further.

BIBLIOGRAPHY

BENHAM, W. E., and HARRIS, I. A., *U.H.F. Performance of Receiving Valves*, Macdonald (1957).

Chapter 6

WAVES ON ELECTRON BEAMS

6.1 The Quantities that Describe a Beam

In the majority of the valves discussed in the last chapter the transit time of the electrons between electrodes is short compared with a r.f. cycle, so that the electrons can be regarded as moving under the influence of fields which remain appreciably constant, and it was seen that it is the gradual departure from this condition as the frequency is increased that leads to the decreasing efficiency and usefulness of these valves.

The principal quality that distinguishes microwave valves from conventional valves is that the condition above is no longer true. In microwave valves the electrons are allowed to take times which are quite long—possibly as many as a hundred r.f. cycles—in transit, and use is made of the way in which disturbances in the electron beam change as this happens.

A beam of electrons can be described by several quantities, such as its charge density, diameter, current, current density, voltage and velocity. Of these, some cannot be defined independently. The current I for instance is given by

$$I = \pi r^2 J \qquad (6.1)$$

where r is the beam radius and J the current density, while

$$J = \rho v \qquad (6.2)$$

where ρ is the charge density and v is the velocity which, in turn, is given as in Chapter 2 by

$$v = (2\eta V)^{\frac{1}{2}} \qquad (6.3)$$

In describing the behaviour of an electron beam we shall use the charge density and velocity. This leaves only the diameter undefined, and this will not matter, since for simplicity we suppose that there is no variation in any direction but that in which the beam is going.

6.2 Velocity and Density Modulation

Consider, then, a d.c. electron beam, with uniform charge density and uniform velocity. Suppose that the beam is passed through a pair of grids which are close together, and across which there is an alternating electric field as in Fig. 6.1. Electrons which pass the

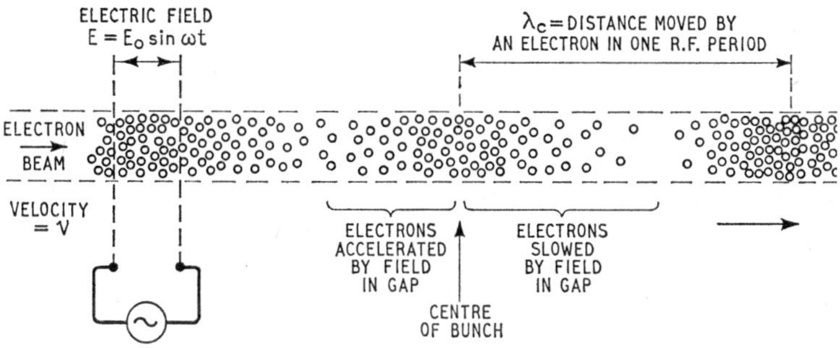

Fig. 6.1. *Density and velocity modulation produced by gridded gap*

grids when the force due to this field is in the direction of motion will be accelerated during their transit, and will emerge with an increased velocity, and those which pass when the field is in the opposite direction will have their velocity decreased, so that if we were able to examine the beam at a plane immediately after leaving the grids, we should find that the charge density would still be constant, but the velocity would be fluctuating with time at the frequency of the signal applied to the grids. To see what happens to such a beam as it progresses, let us take as a reference an electron which crosses the gridded gap just at the end of a r.f. cycle, when the field is instantaneously zero. Such an electron is, of course, neither accelerated nor decelerated and emerges from the gap with the same velocity as that with which it entered, viz. the d.c. velocity. The electrons immediately ahead of it, however, which crossed the gap during the preceding half-cycle, are all going slower, while the corresponding number behind it will all go faster. All this group of electrons will therefore gradually approach together, until their electrostatic repulsion for each other, i.e. the force due to space charge, brings the process to a halt. If the beam is observed by a stationary observer at the point where this occurs, he will see that it consists of bunches of electrons all going past with the same velocity. The *velocity modulation* present on the beam on leaving the gap has all disappeared and been transformed into *density modulation*. As each electron bunch continues on its way the space-charge forces between the electrons come further into play, and so the electrons in the front part of the bunch are accelerated forward away from it, while those at the back are accelerated backward away from it. This reintroduces velocity variations, and when the beam has drifted on the same distance again the density modulation has completely disappeared, and the original velocity modulation has again returned. The process then starts over again, and as the beam drifts on the disturbance on it appears alternately as density and velocity modulation.

Density modulation on an electron beam can be detected, as will be described later, by passing the beam through a similar pair of grids to those used to provide the velocity modulation, and from the foregoing, if such a detector is moved along the beam away from the modulating grids, the signal detected will be as shown in Fig. 6.2. It is seen that there is a standing wave of current density modulation along the beam, and there is also, although it cannot so readily be detected, a similar standing wave of velocity modulation.

6.3 Waves on a Moving Transmission Line

Now, in transmission lines, and since the beam is carrying a.c. power from one place to another it is behaving as a transmission line, we saw in Chapter 3 that standing waves are obtained from the interference between waves going in opposite directions. If we excite a transmission line at one point, the waves travel in both directions away from that point with a velocity of propagation v. Suppose that the transmission line itself is moving past the point of excitation with a velocity u, which is greater than v, then the velocities of propagation past a stationary observer will be $(u+v)$ for the forward-going wave and $(u-v)$ for the backward wave. Both waves will now be going forward, one faster than the line, and the other slower. Since the waves are going at different velocities, they will have different wavelengths. They are both excited at the same point, so at that point they are both in phase, but as they proceed they will gradually

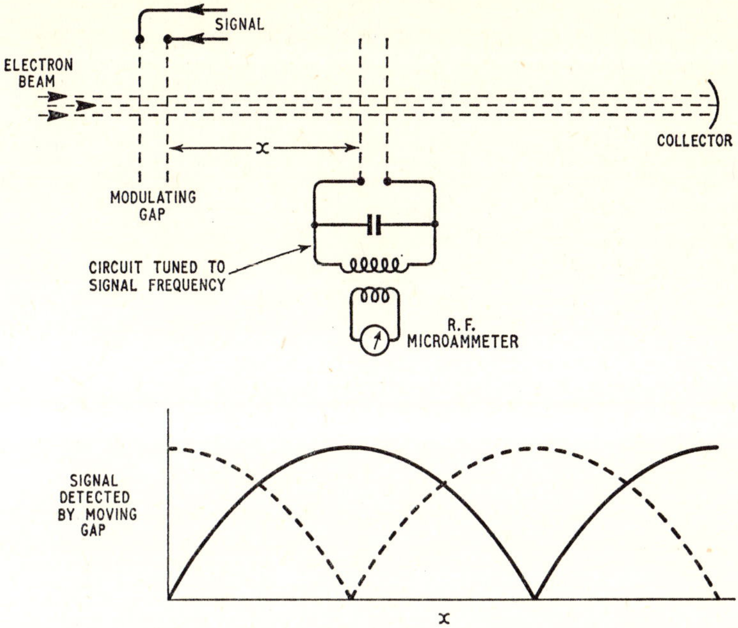

Fig. 6.2. *Density modulation detected on modulated beam. The corresponding velocity modulation is shown by the broken line*

drift alternately out of and into phase, and so a large standing wave will be found along the line as in Fig. 6.3. The *beat-wavelength* of this standing wave (the distance between successive maxima) is given by the condition that the difference in phase between the two waves changes by 2π rad.

For a signal of angular frequency ω, the propagation constant of waves on the stationary line is $\beta_l = \pm\frac{\omega}{v}$; if we invent a new propagation constant $\beta_m = \frac{\omega}{u}$, then the propagation of waves on the moving line, which is described by

$$\beta = \frac{\omega}{u \pm v}, \quad v \ll u \tag{6.4}$$

becomes

$$\beta = \beta_m \left(1 \pm \frac{v}{u}\right) = \beta_m \left(1 \pm \frac{\beta_m}{\beta_l}\right) \tag{6.5}$$

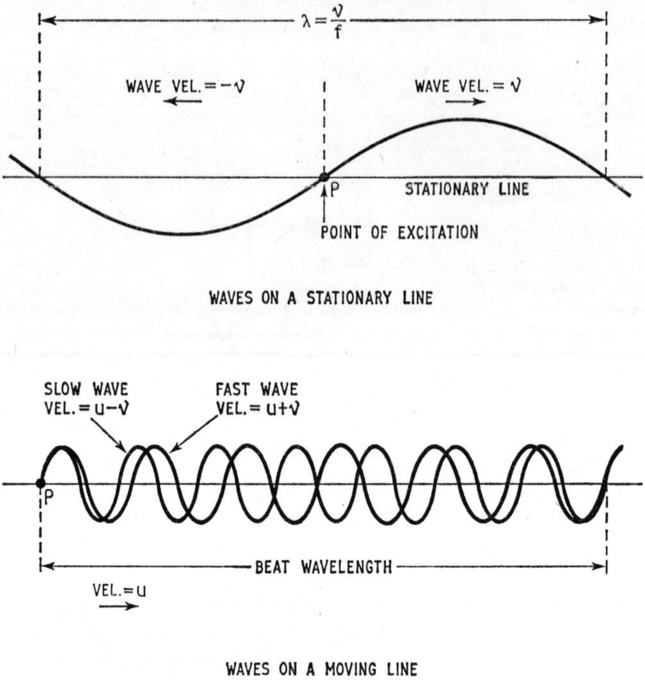

Fig. 6.3. *Waves on stationary and moving lines*

6.4 Waves on a Moving Beam—Plasma Frequency

This state of affairs is somewhat similar to that of the propagation of waves on a moving *electron gas* or *electron plasma*, by which is meant a space filled with electrons. It is not, however, exactly the same, and we must be careful not to push the analogy too far. The elastic behaviour of a stationary electron plasma was first studied by Langmuir, who showed that if the space charge is perturbed by displacing a small volume of it, it oscillates at a frequency $f_p (= \omega_p/2\pi)$ called the *plasma frequency*, which is given by

$$\omega_p^2 = -\frac{\eta\rho}{\epsilon_o} \tag{6.6}$$

where $\eta = e/m$ as before, ϵ_o is the permittivity of free space and ρ is the charge density.

In these oscillations the space charge is being alternately compressed and rarefied, the mechanism being that the initial disturbance of the space charge provides a restoring force, and the inertia of the electrons as they move back causes them to pass through the unperturbed position to give a similar and opposite disturbance by the time they come to rest. The frequency of these oscillations as a function of current density for various beam voltages is shown in Fig. 6.4. This behaviour is also shown by a finite cylindrical electron

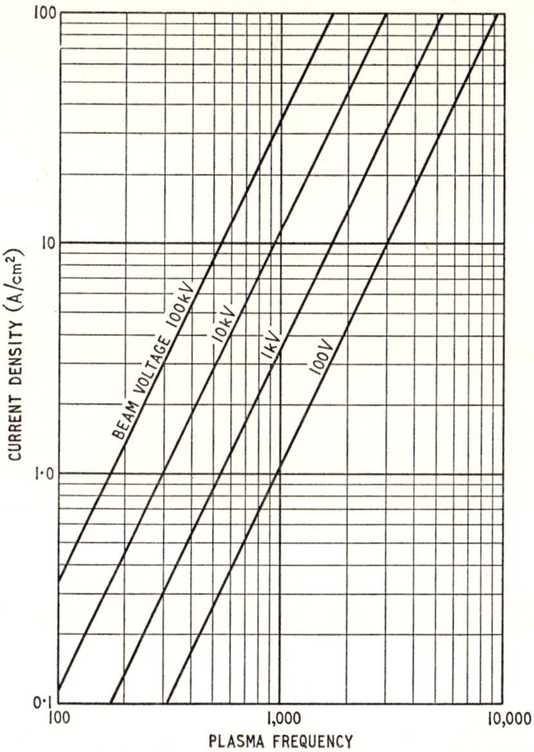

Fig. 6.4.

beam, the only difference being that the frequency of oscillation is reduced due to the different boundary conditions.

It is thus seen that whereas a stationary transmission line has a characteristic propagation velocity, a stationary electron beam has a characteristic oscillation frequency. If a moving electron beam is disturbed in this way as it passes a given point, for instance by being passed through a gridded gap, as in Fig. 6.1, the disturbance will be carried along by the resulting oscillations at the plasma frequency on the beam, and so could be detected at points downstream from the point of disturbance. Suppose the disturbance applied to the beam is at an angular frequency ω, and that the phase velocity of the resulting waves which propagate on the beam is v, then the velocity with which these waves would pass an observer moving with the beam would be $u_0 - v$ or $v - u_0$, and so the number of waves per second passing him would be reduced, and the frequency he would observe would be

$$\frac{\omega}{2\pi} \cdot \frac{v - u_0}{v} \text{ or } \frac{\omega}{2\pi} \cdot \frac{u_0 - v}{v} \tag{6.7}$$

(This is known as the Doppler effect, and in the case of acoustic waves accounts for the apparent change in note from a car horn or train whistle as it passes an observer.)

Since the beam can only carry oscillation at its characteristic plasma angular frequency ω_p, it follows that the wave must propagate at such a velocity that ω_p is the frequency seen by the moving beam, i.e.

$$\omega_p{}^2 = \omega^2 \left(\frac{v - u_0}{v}\right)^2 \tag{6.8}$$

the squares being introduced to embrace both possibilities of $v > u_0$ and $v < u_0$. Then

$$\pm \frac{\omega_p}{\omega} = 1 - \frac{u_0}{v}$$

and hence

$$v = \frac{u_0}{\left(1 \pm \frac{\omega_p}{\omega}\right)} \tag{6.9}$$

6.5 Space-Charge Waves

There are thus two waves which this beam can propagate at frequency ω, one going rather faster than the beam, and the other rather slower, which are therefore referred to respectively as the *fast and slow space-charge waves.*

Equation 6.9 may alternatively be written as

$$\begin{aligned} \beta_{fast} &= \beta_e - \beta_p \\ \beta_{slow} &= \beta_e + \beta_p \end{aligned} \tag{6.10}$$

where $\beta_e = \omega/u_0$ is called the electronic propagation constant and $\beta_p = \omega_p/u_0$.

Another interesting and important point which emerges from a fuller analysis is that in the fast wave, the fluctuations of charge density and velocity are in phase while in the slow wave they are out of phase. The situation is illustrated in Fig. 6.5.

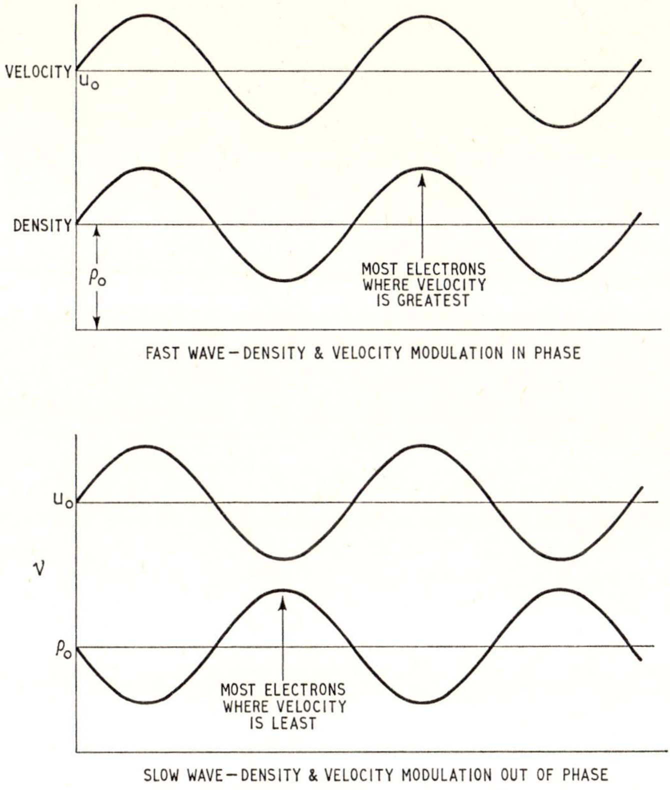

Fig. 6.5. *Velocity and density modulation in fast and slow electron-beam waves*

The significance of this can be appreciated by considering the total energy of a beam carrying a fast space-charge wave. In the regions where the velocity is greater than the d.c. velocity the electron density is also higher than the d.c. level, and hence the total kinetic energy of the beam is increased. This implies that such a wave can be excited on a beam only by supplying energy to it.

The opposite is the case with the slow wave; the greatest number of electrons are now at the points of minimum velocity, and it is consequently found that the energy of a beam carrying a slow space-charge wave is less than that of the d.c. beam, and that such a wave can only be excited on a beam by removing r.f. energy from it. This is not so unreasonable as it may sound. It means that if some of the energy of an electron beam is extracted as r.f. energy, which is what happens in most microwave valves, the resultant modulation on the beam will be principally that of a slow wave.

Because of these characteristics the fast and slow space-charge waves are often spoken of as carrying positive and negative energy respectively, it being understood that these terms are relative to the d.c. energy.

6.5.1 Analysis of Space-Charge Waves

For readers who wish to follow through the mathematics, and as a relatively simple example of the approach used, the following is a derivation of space-charge wave quantities.

Firstly the problem is simplified by supposing that all the beam quantities J, u, ρ, etc, vary only in the direction of the beam velocity, i.e. as if the beam were part of an infinite beam. Next it is assumed that there will be waves, i.e. that the solution we are looking for does exist, and that these waves will propagate as described by the real or imaginary part of $\exp j\,(\omega t - \beta z)$ where β is the unknown propagation constant it is desired to find. This assumption allows $\partial/\partial z$, the partial rate of change with distance in the z direction, which is taken as the direction of the beam, to be replaced by $-j\,\beta$, and $\partial/\partial t$, the partial rate of change with time, to be replaced by $j\omega$.

There are then four equations which can be written: the first, the *continuity of current*, comes from the relation

$$\text{current density } (J) = \text{charge density } (\rho) \times \text{velocity } (u)$$

If all the beam quantities are taken as consisting of a d.c. part plus a relatively small alternating part, e.g.

$$\begin{aligned} J &= J_o + J_\sim \\ u &= u_o + u_\sim \text{ etc.} \end{aligned} \qquad (6.11)$$

then

$$\begin{aligned} J_o + J_\sim &= (\rho_o + \rho_\sim)\,(u_o + u_\sim) \\ &= \rho_o u_o + \rho_o u_\sim + u_o \rho_\sim + \rho_\sim u_\sim \end{aligned} \qquad (6.12)$$

But

$$\begin{aligned} J_o &= \rho_o u_o \\ \therefore \quad J_\sim &= \rho_o u_\sim + u_o \rho_\sim + \rho_\sim u_\sim \end{aligned} \qquad (6.13)$$

However, the quantity $\rho_{\sim}u_{\sim}$ is the product of two small alternating terms, and is therefore negligible compared with terms containing the much larger d.c. components, hence we take

$$J_{\sim} = \rho_0 u_{\sim} + u_0 \rho_{\sim} \tag{6.14}$$

The second equation expresses the *continuity of charge*, i.e. that the rate of change of current density with distance plus the rate of change of charge density with time in the time taken for the beam to go that distance must together be zero, i.e.

$$\frac{\partial J}{\partial z} + \frac{\partial \rho}{\partial t} = 0 \tag{6.15}$$

It is defined that only the alternating components are varying, so that by applying

$$\frac{\partial}{\partial z} = -\mathrm{j}\beta, \quad \frac{\partial}{\partial t} = \mathrm{j}\omega$$

this becomes

$$\mathrm{j}\omega\rho_{\sim} - \mathrm{j}\beta J_{\sim} = 0$$

or

$$J_{\sim} = \frac{\omega}{\beta}\rho_{\sim} \tag{6.16}$$

Thirdly, there is *Poisson's equation*, which expresses the relation that the divergence of electric displacement ($\epsilon_0 E$ where E is the electric field strength) is proportional to the charge density. In this one-dimensional picture this becomes

$$\epsilon_0 \frac{\partial E}{\partial z} = \rho \tag{6.17}$$

or, inserting the d.c. and alternating components,

$$\epsilon_0 \left[\frac{\partial E_0}{\partial z} + \frac{\partial E_{\sim}}{\partial z}\right] = \rho_0 + \rho_{\sim} \tag{6.18}$$

In this the time-varying and non-time-varying components must be equal on both sides of the equation, so that

$$\epsilon_0 \frac{\partial E_{\sim}}{\partial z} = \rho_{\sim}$$

i.e.

$$-\mathrm{j}\beta E_{\sim} = \frac{\rho_{\sim}}{\epsilon_0} \tag{6.19}$$

Lastly there is the *force equation* for an electron in the field E,

$$eE = -\frac{\mathrm{d}(mu)}{\mathrm{d}t} \tag{6.20}$$

Replacing the total differential $\mathrm{d}/\mathrm{d}t$ by the sum of its two partial components

$$\frac{\mathrm{d}}{\mathrm{d}t} = \frac{\partial}{\partial z}\frac{\partial z}{\partial t} + \frac{\partial}{\partial t}$$

Separating out the alternating components, and putting $\eta = \frac{e}{m}$ this becomes

$$\eta E_{\sim} = -\left(\frac{\partial u_{\sim}}{\partial z}\frac{\partial z}{\partial t} + \frac{\partial u_{\sim}}{\partial t}\right) \tag{6.21}$$

and again using

$$\mathrm{j}\beta = -\frac{\partial}{\partial z}, \quad \mathrm{j}\omega = \frac{\partial}{\partial t} \quad \text{and} \quad u = \frac{\partial z}{\partial t}$$

$$\eta E_{\sim} = \mathrm{j}\beta u_o u_{\sim} - \mathrm{j}\omega u_{\sim} \tag{6.22}$$

Now $E_{\sim}$ can be eliminated from Equation 6.19 and 6.22 giving

$$\frac{\eta \rho_{\sim}}{\beta \epsilon_o} = -u_{\sim}(\omega - \beta u_o) \tag{6.23}$$

and $J_{\sim}$ can be eliminated from Equations 6.14 and 6.16, giving

$$(\omega - \beta u_o)\,\rho_{\sim} = \beta \rho_o u_{\sim} \tag{6.24}$$

Dividing Equation 6.23 by Equation 6.24:

$$\frac{\eta}{\beta \epsilon_o(\omega - \beta u_o)} = \frac{-(\omega - \beta u_o)}{\beta \rho_o}$$

or, rearranging

$$\frac{\eta \rho_o}{\epsilon_o} = -(\omega - \beta u_o)^2 \tag{6.25}$$

We recognize $-\eta\rho_o/\epsilon_o$ as $\omega_p{}^2$, the characteristic angular frequency of vibration of an electron plasma of charge density ρ_o, so Equation 6.25 becomes

$$\omega_p{}^2 = (\omega - \beta u_o)^2 \tag{6.26}$$

We have now eliminated all the unknown a.c. quantities from the equations, and can solve for β, to find how waves of angular frequency ω propagate on the beam.

We have

$$\omega - \beta u_o = \pm\, \omega_p$$

hence

$$\beta = \frac{\omega}{u_o} \pm \frac{\omega_p}{u_o} \tag{6.27}$$

which is the result given earlier. It is seen that two waves are possible, a slow wave, having

$$\beta_s = \frac{\omega}{u_o} + \frac{\omega_p}{u_o} \tag{6.28(a)}$$

and a fast wave

$$\beta_f = \frac{\omega}{u_o} - \frac{\omega_p}{u_o} \tag{6.28(b)}$$

Also, from Equations 6.14 and 6.16 we get

$$J_{\sim}\left(1 - \frac{\beta u_o}{\omega}\right) = \rho_o u_{\sim} \tag{6.29}$$

Substituting the two values of β obtained in Equation 6.28 gives

$$J_{\sim f} = \frac{\omega}{\omega_p}.\frac{J_o}{u_o}.\, u_{\sim f} \tag{6.30(a)}$$

for the fast wave, and

$$J_{\sim s} = -\frac{\omega}{\omega_p}\frac{J_o}{u_o}\, u_{\sim s} \tag{6.30(b)}$$

for the slow wave. These two equations give the amount of current modulation that appears on a beam for a corresponding amount of velocity modulation. The minus sign of Equation 6.30(b) shows that in the slow wave the current and velocity are in opposite phase, as stated earlier, while in the fast wave they are in phase.

BIBLIOGRAPHY

SIMS, G. D., and STEPHENSON, I. M., *Microwave Tube and Semiconductor Devices*, Blackie and Son (1963).

BECK, A. H. V., *Space Charge Waves and Slow Electromagnetic Waves*, Pergamon Press (1958).

HUTTER, R., *Beam and Wave Electronics in Microwave Tubes*, Van Nostrand (1960).

REICH, H. J., ORDNUNG, P. F., KRAUSS, H. L., and SKALNIK, J. G., *Microwave Theory and Techniques*, Van Nostrand (1953).

Chapter 7

VALVES WITH LOCALISED INTERACTION

7.1 Gridded Gap with Finite Transit Time

Conventional valves, such as those dealt with in Chapters 4 and 5, all make use of localised interaction and have their interacting regions, e.g. the cathode-grid space and the grid-anode space, immediately adjoining. Some microwave valves also make use of localised interaction, but have the regions of interaction relatively widely separated, while others, such as the travelling-wave tube and magnetron make use of continuous interaction between a wave and a beam. The two types are rather different in conception, and because of their logical development from conventional valves, it will be convenient to deal with the former type first. In order to do this, it will first be necessary to consider in a little more detail the interaction of a beam with a gridded gap across which a r.f. field is being applied.

As we have seen, if the transit time across the gridded gap is small compared with the period of a r.f. cycle, and the whole system has an infinite cross-section so that there is no radial variation, then there is no difficulty, and the effect of the gap is to impose a velocity variation on the emergent beam. We may now notice that some electrons will be accelerated as they cross the gap, but an equal number in the following half-cycle will be retarded, and on average no energy will be supplied by the field. Consequently both slow and fast waves must be excited with equal amplitudes.

Suppose now that the two grids are moved gradually farther

apart; what is the effect on the result? The field will now be varying as the electron passes through the gap, and so the result must be obtained in the simplest case, neglecting space-charge effects, by integrating the force equation

$$\frac{\mathrm{d}}{\mathrm{d}t}(mu) = -eE \sin \omega t \tag{7.1}$$

over the period taken for the electron to pass. This differential equation is of a simple kind, which can be integrated directly, giving

$$\int \mathrm{d}(mu) = \int_t^{t+t'} -eE \sin \omega t \mathrm{d}t$$

The left-hand side then gives the total change in the momentum, $mu - mu_o$, while the right-hand integration, carried out from the time t when the electron enters the gap, to $t + t'$ when it leaves, gives

$$m(u - u_o) = \frac{eE}{\omega}[\cos \omega(t + t') - \cos \omega t]$$

Since $u = u_o + u_\sim$, using a standard trigonometrical relation we obtain

$$u_\sim = 2\frac{\eta E}{\omega} \sin \frac{\omega t'}{2} \sin \omega \left(t + \frac{t'}{2}\right) \tag{7.2}$$

Now, the time of transit t' is usually expressed by the transit angle θ, which is defined by $\theta = \omega t'$. With a gap width d and a peak r.f. gap voltage V_1, we have the simple relations

$$V_1 = Ed, \quad u_o t' = d \text{ and } u_o = (2\eta V_o)^{\frac{1}{2}}$$

so that $u_\sim$ can be expressed as

$$u_\sim = \left(\frac{\eta}{2V_o}\right)^{\frac{1}{2}} V_1 \frac{\sin \frac{\theta}{2}}{\frac{\theta}{2}} \sin\left(\omega t + \frac{\theta}{2}\right) \tag{7.3}$$

This equation gives the amount of velocity modulation produced on a beam of d.c. voltage V_o by the r.f. voltage V_1 applied across a gap of transit angle θ. The last term, $\sin\left(\omega t + \frac{\theta}{2}\right)$ gives the phase of the velocity modulation relative to the applied signal, which was given by $\sin \omega t$. As might be expected, the modulation lags on the applied voltage by half the transit angle.

The function sin $\frac{1}{2}\theta/\frac{1}{2}\theta$ appearing in the expression shows that the efficiency of modulation falls off as the gap width is increased, and is called the *gap factor*. This function is one which often turns up when discontinuous processes are considered and is shown plotted in Fig. 7.1. It can be seen that, although at first the effect of widening the gaps is not serious, as soon as the transit angle exceeds about 1 rad the gap factor becomes more and more significant, and ultimately completely spoils the interaction.

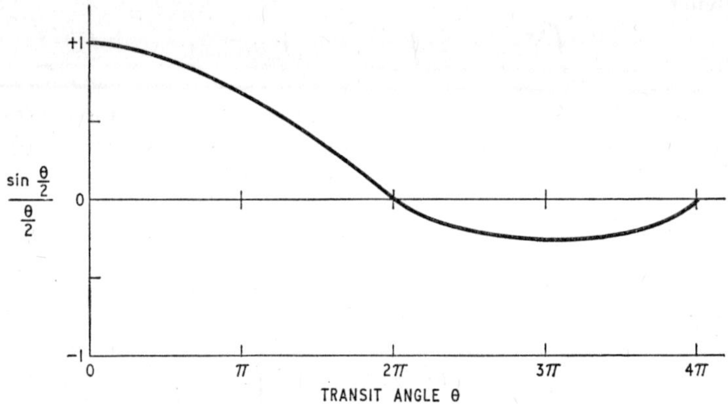

Fig. 7.1. *Variation of gap factor with transit angle*

The other finite dimensions which have not been considered are the thickness of the beam and the size of the holes in the grid. In a beam of finite thickness there is always some velocity reduction as the centre is approached, due to the space-charge fields produced by the electrons. These we shall continue to neglect, but we may note for justification that they can be and often are taken into account by considering the beam as being made up of several beams of slightly different velocities. The other effect is usually more serious, because although we have spoken of a pair of grids, assumed not to intercept any electrons, it is rarely possible to use such grids at very high frequencies and high power levels, and the physical arrangement is more often as shown in Fig. 7.2, where the beam crosses the field region between the ends of two cylinders, or the 'lips' of a resonant cavity. The efficiency of such an arrangement depends on two ratios, firstly hole diameter/wavelength and secondly beam diameter/hole diameter. The first ratio is usually expressed as a quantity *ka*, where

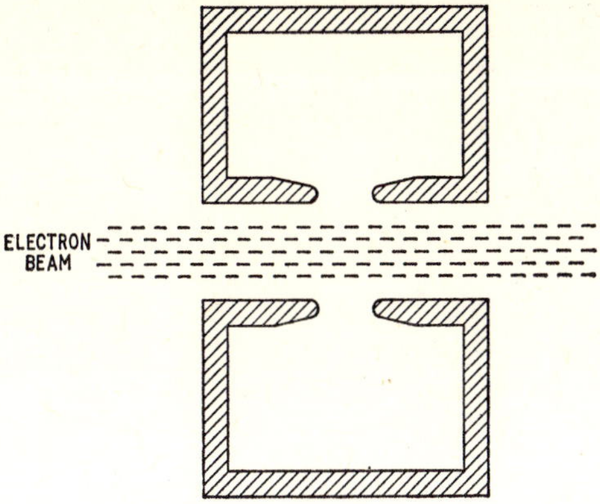

Fig. 7.2. *Resonant cavity gap*

$k = 2\pi/\lambda$, a is the hole radius, and λ is the wavelength, and the second by the ratio b/a where b is the beam radius.

The complete interaction in a gap then depends on the transit angle (i.e. the gap width), the gap diameter and the extent to which the beam fills the hole, and each of these factors can reduce the efficiency below that of the ideal of a very narrow gap in which both grids consist of very fine wires which do not intercept any current. We shall, however, continue to use this latter concept, remembering that in practice things are not so convenient.

7.2 Two-Cavity Klystrons

The principal family of microwave valves in which this type of interaction is used are the *klystrons*, of which we shall first consider the two-cavity klystron, invented by the brothers Varian in 1939. The klystron is one example of the practice of giving every new type of variant of thermionic device a name ending in 'tron'. This particular name is derived from the Greek word for the breaking of a wave.

As may be seen from the diagram of Fig. 7.3, the klystron consists of a thermionic cathode C from which electrons are drawn to form a beam by an electrode A, connected to a higher voltage. This electrode is often referred to as an anode, but is more correctly called a grid. The unmodulated beam is shot through the aperture of a resonant cavity B, which is being fed with a small signal at its resonant frequency. The peak fields generated across the cavity

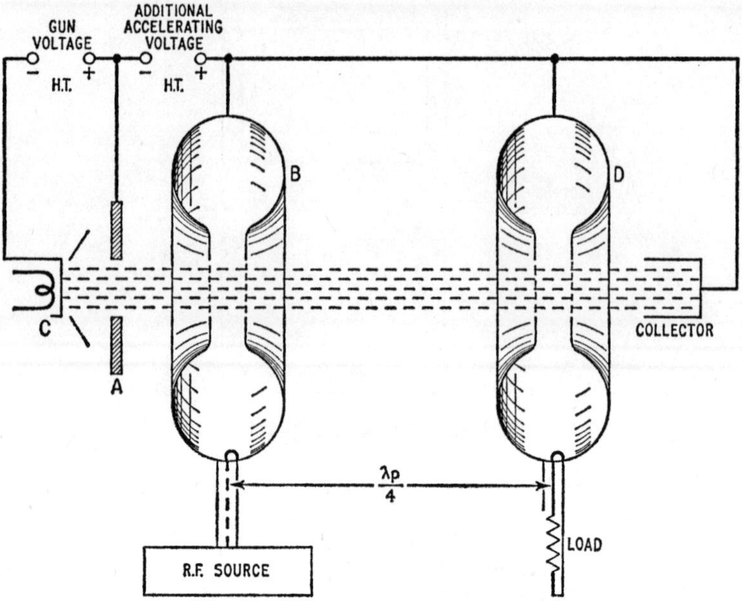

Fig. 7.3. *Two-cavity klystron amplifier*

aperture produce a velocity modulation on the beam, as described in the last chapter, and the beam is then allowed to drift for the distance, one-quarter of the plasma wavelength, in which this velocity modulation is converted by the interference between the two space-charge waves into a maximum of density modulation.

As has been shown, the velocity modulation which is impressed on the beam is given by Equation 7.3, which in terms of u_o becomes

$$u_{\sim} = \frac{\eta V_1}{u_o} . A_1 \frac{\sin \frac{\theta}{2}}{\frac{\theta}{2}} \sin \left(\omega t + \frac{\theta}{2}\right) \qquad (7.4)$$

where A_1 is a constant to take into account radial variations (i.e. the effect of the hole and beam diameters). In the last chapter it was seen that two waves can propagate along a beam. In these two waves the relation between current modulation and velocity modulation is given respectively by

$$I_{\sim f} = \frac{\omega}{\omega_p} \frac{I_o}{u_o} u_{\sim f}$$

$$I_{\sim s} = -\frac{\omega}{\omega_p}\frac{I_o}{u_o}u_{\sim s}$$

as in Equation 6.30. The total velocity modulation, and the total current modulation at any point must be given by the sum of these two waves. At the input cavity, we have $u_{\sim total}$ as given in Equations 7.4 and $I_{\sim total} = 0$.

It follows therefore that the two waves are excited in equal amplitudes, so that at this point the in-phase current modulation of the fast wave is exactly cancelled at all times by the out-of-phase current modulation of the slow wave.

The two space-charge waves thus excited will then propagate sinusoidally along the beam, so that

$$u_f = \tfrac{1}{2} u_\sim \sin(\beta_e + \beta_p) z$$

and

$$u_s = \tfrac{1}{2} u_\sim \sin(\beta_e - \beta_p) z \qquad (7.5)$$

and their sum

$$u_{total} = u_\sim \sin\beta_e z \cos\beta_p z \qquad (7.6)$$

At the point where $\beta_p z = \pi/2$, which is when $z = \lambda_p/4$, the two velocity modulations will always be out of phase, and the density and current modulations will therefore always be in phase. This is shown diagrammatically in Figs. 6.3 and 6.4. The current modulation at this point will then be

$$I_\sim = \frac{\omega}{\omega_p}\frac{I_o}{u_o}u_\sim$$

$$= \frac{\omega}{\omega_p}\frac{I_o}{u_o^2}\eta A_1 \left(\frac{\sin\frac{\theta_1}{2}}{\frac{\theta_1}{2}}\right) V_1 \qquad (7.7)$$

or, remembering that $u_o^2 = 2\eta V_o$,

$$I_\sim = \frac{\omega}{\omega_p}\frac{I_o}{2V_o}K_1 V_1 \qquad (7.8)$$

where K_1 $(=A_1 \sin\frac{1}{2}\theta_1/\frac{1}{2}\theta_1)$ represents the departure of the input cavity from the ideal.

When the electron bunches arrive at the second cavity, which is taken for the moment as being resonant at the input frequency, the cavity will, in the steady-state condition, be resonating so that the field across its gap is at a maximum and opposing the electron flow. If the peak instantaneous voltage across the gap is V_2, then the mean energy W_2 given up by the beam will be $I_{\sim}V_2/2$, or again since similar gap and aperture factors will apply, applying Equation 7.8

$$W_2 = \frac{\omega}{\omega_p} \frac{I_o}{4V_o} K_1 K_2 V_1 V_2 \qquad (7.9)$$

This is the power extracted from the beam, some of which power is dissipated in the loss of the output cavity, while the remainder is given up to an external load. Since, in principle, no energy is supplied to the beam by the input cavity, the only power supplied from the r.f. drive is that required to make up the loss of the input cavity; hence for maximum gain it is clearly desirable to have this loss as small as possible. It is also necessary to make V_1 and V_2 as large as possible. Both of these require that the Q's of the two cavities should be as large as possible.

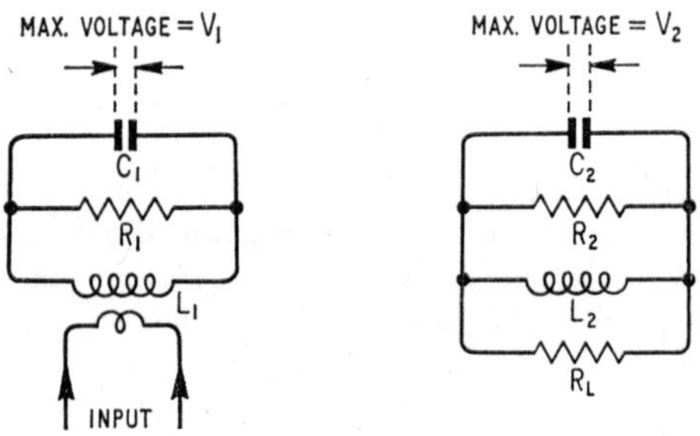

Fig. 7.4. *Klystron equivalent-circuits*

If the equivalent circuits shown in Fig. 7.4 are taken as representing the input and output cavities, then the mean power dissipated in the resistance R_1 of the input cavity is

$$W_1 = \frac{V_1^2}{2R_1}$$

and this is therefore the input power to the valve. The output power from the beam, as given above, can then be expressed as

$$W_2 = \frac{\omega}{\omega_p} \frac{I_o}{4V_o} K_1 K_2 V_2 (2R_1 W_1)^{\frac{1}{2}} \qquad (7.10)$$

This is the power dissipated in the losses in the output cavity, $V_2^2/2R_2$, and in the external load $V_2^2/2R_L$, and is, therefore,

$$W_2 = \frac{V_2^2}{2}\left(\frac{1}{R_L} + \frac{1}{R_2}\right)$$

hence

$$\frac{V_2}{2}\left(\frac{1}{R_L} + \frac{1}{R_2}\right) = \frac{\omega}{\omega_p}\frac{I_o}{4V_o} K_1 K_2 (2R_1 W_1)^{\frac{1}{2}} \qquad (7.11)$$

The power actually delivered up to the external load is

$$W_L = \frac{V_2^2}{2R_L}.$$

Inserting the value of V_2 from this in Equation 7.11, squaring both sides and rearranging, we get for the overall gain

$$\frac{W_L}{W_1} = \frac{4R_L R_1}{\left(1 + \frac{R_L}{R_2}\right)^2}\left(\frac{\omega}{\omega_p}\cdot\frac{I_o}{4V_o}\right)^2 (K_1 K_2)^2 \qquad (7.12)$$

This implies, as would be expected, that for maximum gain, R_1 and R_2 should be as large as possible, which is another way of saying that the Q's of the input and output cavities should be a maximum, and that K_1 and K_2, both necessarily less than one, should also be as large as possible, i.e. ideally very short gridded gaps should be used.

The quantity $I_o/4V_o$ has the dimensions of an impedance and is sometimes called the beam impedance. If we take the whole middle bracket of Equation 7.12, $\left(\frac{\omega}{\omega_p}\cdot\frac{I_o}{4V_o}\right)^2$, remembering that

$$\omega_p^2 = \frac{-\eta\rho}{\epsilon_o}$$

and

$$I_o = \rho u_0 = \rho\,(2\eta V_o)^{\frac{1}{2}}$$

it is clear that these two terms are related, and it actually becomes the quantity ρ/V_o that should be maximised.

After passing through the output cavity, the electron beam, now depleted in energy, passes on to a collector which may be externally cooled to dissipate the waste energy, which appears as heat.

7.3 Energy Exchange

The general principle used here, as has already been mentioned in Chapter 4 in connection with conventional valves, is the same as is used in all valves in which d.c. energy is converted into r.f. energy, namely that the electrons are so organised that they only interact with the r.f. fields when those fields are in the right direction to extract energy from them. In this case, as with the triode, this is done automatically, since matters arrange themselves so that electrons pass through the field gap predominantly when the r.f. field opposes their motion, while only a reduced number, ideally none, pass when the field assists their passage, and so gives up energy in accelerating them.

7.4 Comparison with a Triode

It may be instructive to take this comparison with a triode a little farther, since it will be seen that, apart from the drifting-beam phenomenon, a klystron of this sort is actually very much simpler in its operation than is a triode, since all its functions are separated. In a triode the mean beam current is controlled by the mean potential applied to the grid. The beam bunching is then brought about directly by the r.f. fields applied across the cathode-grid space. The bunches pass directly into the second r.f. interaction gap, the grid-anode space, being phased as before so that the majority pass when the r.f. field opposes their passage, and are at the same time accelerated by the d.c. field also applied across this gap. In the klystron all these functions are separated. The beam current is controlled by the electrode A, the beam is then separately accelerated, modulated by the first gap followed by the drift space, and shot into the second interaction gap which again is free from d.c. fields.

It will be appreciated that since the distance between the cavities is one-quarter plasma wavelength, i.e. the beat wavelength between the two space-charge waves, the transit time of electrons between the two cavities will be several times the period of the radio frequency being amplified. It is now only the actual transit times through the gaps which it is necessary to keep short, and this it is relatively easy to do, since the beam velocity can be made quite high.

7.5 Properties and Use of Two-cavity Klystrons

The device which has been described is clearly a narrow band amplifier, which might be tuned by tuning the resonant frequencies of both cavities simultaneously. The higher the Q-factors of the two cavities can be made, the higher will be the gain which can be obtained, but, of course, the narrower the bandwidth will become;

also the less the coupling to the output cavity, the less will be the power that can be extracted relative to the loss in the cavity.

Like any other amplifier, the two-cavity klystron can readily be turned into an oscillator, by feeding back some of the power from the output cavity to the input cavity through a short coaxial line, and the earliest klystron oscillators usually worked in this way. Owing to the need for careful adjustment of both cavities and the feedback circuit, such oscillators were difficult and touchy things to use, and this function is now usually filled by the reflex Klystron, which will be described shortly.

The chief modern use for two-cavity klystrons is as power amplifiers, particularly where it is desired to conserve only a relatively narrow bandwidth and to avoid amplifying any noise at frequencies near by. This is a requirement which arises, for instance, in a c.w. Doppler radar, where the frequency of the signal reflected from a moving target will be slightly altered by the Doppler effect, and can be compared directly with the transmitted frequency. The limit to the sensitivity of such a system is the noise being transmitted at the shifted frequency, and it will be clear from the above that a klystron characteristic is very suitable for such an application. Such valves are usually made with the cavities and body brazed together, forming the vacuum envelope, and will be watercooled if the powers involved are high enough to make this necessary. They also need some means of keeping the beam focused as it passes through the valve, and this is done by external solenoid windings or a permanent magnet. An example of such a valve is shown in Plate 7.1.

7.6 Multi-Cavity Klystrons

If more gain or more bandwidth is required than can be obtained from the use of two cavities, the *gain-bandwidth product* can be increased by interposing further cavities between the input and the output cavities. These cavities are usually completely unloaded, i.e. no power is coupled out of them, and so they oscillate at high amplitude at the frequency at which they are being driven by the beam, and in so doing they modify the beam bunching. The net action is rather like that of a stagger-tuned bandpass circuit.

Multi-cavity klystrons provide the highest powers that are at present obtainable at microwave frequencies, coupled with very high gains which are advantageous, since they then avoid the need for relatively high-power valves as drivers. As an indication of the power and frequency range which they cover, valves have been made giving over 30 MW pulsed power at 1,000 Mc/s and over 100 kW c.w. at about 10 Gc/s.

The construction of such valves varies according to the frequency at which they are to be used. At frequencies above about 1,000 Mc/s the cavities are sufficiently small so that they can reasonably be made part of the valve. They are then brazed together, and form part of the vacuum envelope. They are provided with liquid cooling, and mechanically tuned by having movable elements mounted on metal bellows to conserve the vacuum seal. The construction of a klystron of this type, which is an amplifier producing an output of 3 kW at 10 Gc/s, with a bandwidth of 4·5 Mc/s is shown in Plate 7.2. At lower frequencies one of the chief uses of multi-cavity klystrons is as u.h.f. television transmitters, where they are required to produce powers of up to 50 kW at frequencies in the 500 to 1,000 Mc/s range, with a bandwidth of about 5 Mc/s. At these lower frequencies the resonant cavities are too large to be incorporated in the valve, and so the valves are made with short gaps, connected together by ceramic cylinders, and the large external cavities, which are usually of rectangular section, are connected to the edges of the gaps by spring fingers. An example of this type is shown in Plate 7.3.

7.6.1 Efficiency

The maximum efficiency of power klystrons when working at their maximum output is about 45 to 50 per cent, but it frequently happens, as is the case with the u.h.f. television transmitters, that in order to obtain adequately linear operation characteristics, it is necessary to work below the maximum output power and consequently with lower efficiency. This is a characteristic of all electron-beam tubes, and one method of increasing this efficiency, by the use of *collector-depression*, is applicable to these.

7.6.2 Collector Depression

The efficiency is limited, because it is not possible to extract all the kinetic energy of the beam. As soon as the slowest electrons are brought to rest, they cease to travel with the beam, and the resulting increase in charge density usually destroys the beam formation, and excessive interception currents occur at the cavity gaps, often with disastrous results. The beam, therefore, must be allowed to leave the output cavity with a reasonably high velocity. When the valve is being used at less than its maximum output power, the average electron velocity after the output cavity can be quite high, rising right up to the d.c. velocity in the limiting case of a very small signal. There is no reason, however, why the electrons should not be considerably slowed down before they are collected, which can be brought about by decreasing the potential of the collector so that the electrons are decelerated as they approach it. This has the dual advantages that less power is supplied by the power supply, thus increasing the efficiency of operation, and less heat is dissipated in the collector, thus simplifying the cooling problem. One method of arranging the power supplies for this type of operation is shown in Fig. 7.5. In this arrangement the beam current is controlled by

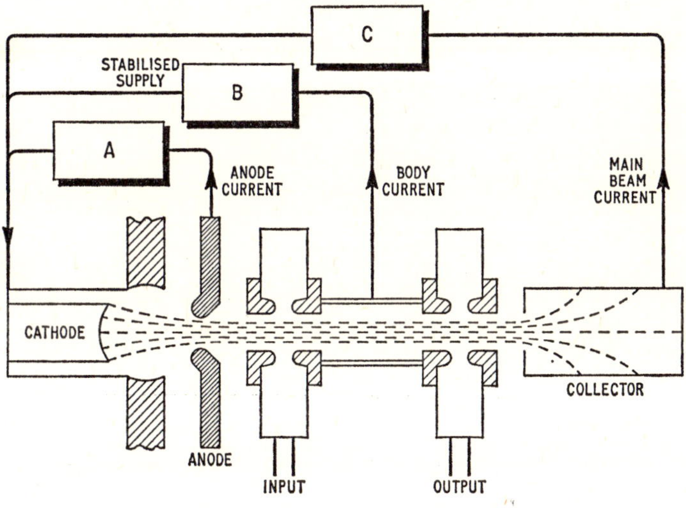

Fig. 7.5. *Arrangement for operation with collector depression*

supply *A*, which has only to supply any current that is intercepted by the grid, which is normally very small. The main high-voltage supply *B* determines the beam velocity through the cavities, and must supply any interception currents flowing to the body. This is typically considerably less than 10 per cent of the total beam current. The relatively small current now supplied by this power supply also makes it easier to provide the fairly high degree of stabilisation that is required. The main beam current through the valve is supplied by the third supply, *C*, which can be as much as 50 per cent below *B* in voltage, and is also much less critical in its stability requirement.

7.7 Reflex Klystrons

Similar in some ways to the two-cavity klystron is the *reflex klystron*, which is shown diagrammatically in Fig. 7.6. In this arrangement an electron beam is drawn as before from the cathode *C* by the grid *G*, and passed through a resonant cavity. The beam then enters a region in which the static retarding field provided by the reflector first brings the electrons to rest, and then accelerates them back

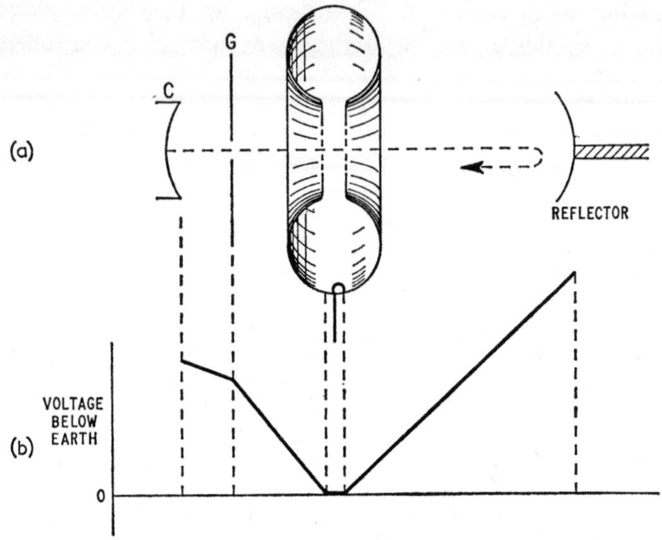

Fig. 7.6. *System of reflex klystron (a) cross-section (b) voltage distribution*

through the cavity gap. The velocity modulation impressed on the beam at the first passage through the r.f. gap again causes bunching to take place during the interval in which the electrons are brought to rest and returned to the gap, and provided that this time interval is of the right length, the bunch will arrive back at the gap in a suitable phase to give up energy by interacting with the r.f. field. This interval can be adjusted by varying the potential on the reflector. The valve is therefore an oscillator, since the feedback is provided directly by the electron beam, and the frequency of oscillation is controlled coarsely by tuning the resonant cavity, or more finely by varying the voltage of the reflector. Oscillations can be obtained for several values of the reflector voltage, as the interval between the forward and return transit through the cavity varies successively by a r.f. period, and the value giving the best output must usually be selected by experiment.

Cavity tuning at the lower frequencies is usually carried out as in multi-cavity klystrons by screwing plugs into the cavity, and an example of an early reflex klystron tuned in this way is shown in Plate 7.4. At higher frequencies the tuning is more often obtained by physically distorting the cavity.

The principal use of reflex klystrons is as low-level oscillators, for example as local oscillators in superheterodyne receivers. The beam powers involved are therefore usually quite small, and no magnetic focusing circuits are necessary. After the second transit through the cavity, the beam is allowed to diverge, and is collected on either the cavity or the grid. The fine tuning provided by the reflector-voltage variation makes the valves particularly suitable as local oscillators, since it provides a means of applying automatic frequency control to cope with any drift of the received signal.

Tuning over the widest frequency ranges, for use in signal generators, is now usually achieved by the use of coaxial cavities, and an example of the best current practice in the 6,000 Mc/s region, in which the valve plugs into an external coaxial cavity, is shown in Plate 7.5.

7.8 Coaxial line Oscillators

Another valve in the present category, which has developed from one of the very earliest proposals in this field by A. and O. Heil, is the coaxial line oscillator, sometimes still referred to as a 'Heil tube'. In the form in which it is now most frequently used it consists of a tunable coaxial cavity, with an electron beam projected across it on a diameter, as indicated in Fig. 7.7. The beam, often of rectangular or strip section, is accelerated and fired through the gap between the inner and outer cylinders of the coaxial line; it then drifts across the field-free interior of the inner cylinder, as usual becoming bunched as a result of the velocity modulation impressed on it by the radial r.f. field across the first gap, and again interacting with this field as it crosses the second gap. From the electron-beam point of view, there is very little difference between this arrangement and that of the two-cavity klystron or the floating-drift-tube klystron to be described next. It is principally in the r.f. circuit arrangements that differences arise, and in the coaxial-line oscillator there is obviously a direct feedback path between the two interaction gaps, so that the

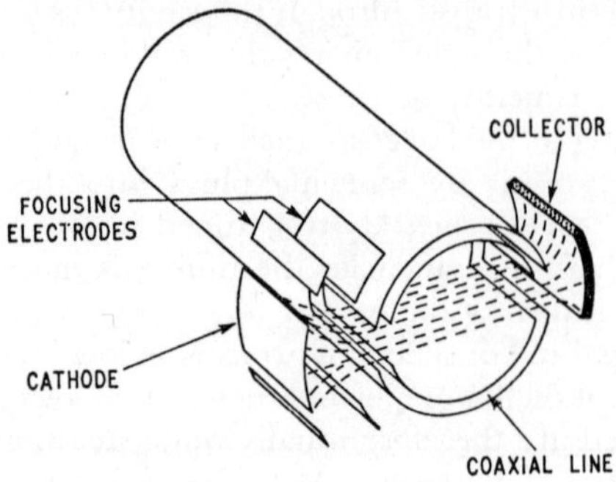

Fig. 7.7. *Coaxial-line oscillator*

device is an oscillator whose frequency is fixed by the resonance of the coaxial line cavity. At one time this valve was outstanding for the very great tuning range which it offered, a range of over two to one in frequency being possible, but the development of coaxial cavities for use with klystrons has now to a large extent evened up this situation. As with the reflex klystron, some extent of electronic tuning can be obtained by variation of the beam velocity, and in addition to their use as oscillators, at low levels of a few milliwatts, these valves are also widely used at power levels of up to about 1 watt as directly modulated r.f. sources for low-capacity radio links.

7.9 Floating Drift-Tube Klystrons

Another variant on the same theme, of having two localised interaction gaps connected by a drift space, is the *floating-drift-tube* klystron. This again resembles the two-cavity klystron, as can be seen from Fig. 7.8, but differs in that, as in the Heil tube, the two gaps are directly connected together by the coaxial region outside the floating drift tube. It is therefore once again an oscillator, but since the beam can pass right through the valve and be collected on a separate collector, it can be designed for much higher powers than the reflex klystron, while it preserves the simplicity of tuning that arises from only having one resonant cavity.

7.10 Harmonic Generation Klystrons

The valves so far described comprise a set of oscillators and amplifiers which extend in frequency from a few hundred mega-

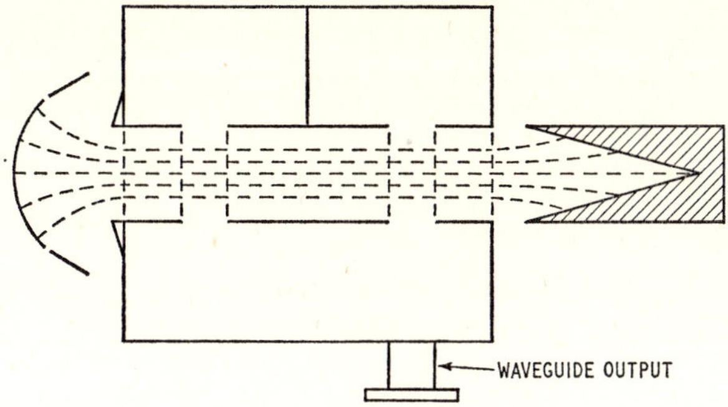

Fig. 7.8. *Floating drift-tube klystron*

cycles to about 60,000 Mc/s, where the free-space wavelength is only 5 mm, in c.w. power from less than 1 mW up to some hundreds of kilowatts and in pulse power to the order of 10 MW. At the very high frequency end of this range the difficulties of manufacture become increasingly great, and one approach that has been taken for the generation of the highest frequencies is to produce bunching on a beam at a conveniently lower frequency, say 35,000 Mc/s, and to use these bunches to interact with the fields in a cavity tuned to a harmonic of this frequency. This means that the higher frequency output cavity is only driven by an electron bunch every second, third, or fourth cycle, but since the cavity has a relatively high Q-factor, and the bunching that can be applied at the lower frequency is quite tight because of the higher drive powers that are available there, it is a worth-while approach. A klystron of this sort is known as a *harmonic-generation* klystron.

As frequency increases, however, losses inevitably become greater, cavity sizes become smaller, difficulties of manufacture increase, and the current densities required in the electron beam, whose diameter steadily decreases, become more and more difficult to obtain. It is, at least at present, difficult to visualise any very great extension of the present frequency range.

BIBLIOGRAPHY

WARNECKE, R., and GUENARD, P., *Tubes a Modulation de Vitesse Klystrons,* Gauthier-Villars (1951).

KLEEN, W., *Electronics of Microwave Tubes,* Academic Press (1958).

HAMILTON, D. R., KNIPP, J. K. and KUPER, J. B. N., *Klystrons and Microwave Triodes,* M.I.T. Radiation Laboratory Series (1948).

Chapter 8

THE HELIX-TYPE TRAVELLING WAVE TUBE

8.1 Travelling Wave Interaction

In the helix-type travelling wave tube we meet for the first time the idea of a drifting electron beam continuously interacting with a travelling r.f. field as it passes through the valve. This principle of continuous interaction is used in both valves in which there are no d.c. electric fields in the interaction region (often called O-type valves) and crossed-field (often called M-type) valves, although their mechanisms of interaction are different; it will again be seen that, as in all other valves, the basic mechanism depends on bunching the electrons so that they are predominantly moving against the forces due to r.f. fields, and so can give up energy to them.

The helix travelling wave tube, like the magnetron, is a deceptively simple-looking device, and any attempts to describe its operation in simple terms are unlikely to be accurate in detail. However, let us first attempt a simple examination; the complicating detail can be added later where necessary. It consists, as indicated in Fig. 8.1, of a helical slow-wave circuit through which an electron beam is projected. If, for simplicity, we suppose the helix to be within an outer conducting cylinder, it forms a two-conductor transmission line, and so will propagate at all frequencies down to d.c. Its velocity of propagation, however, is considerably less than that of a coaxial line, and to a good approximation it behaves as though the r.f. wave were propagating along the wire of the helix at the velocity of light, $\mathbf{c}$. The mean axial velocity, hereafter called the circuit velocity, is given by

$$v = \mathbf{c} \sin \psi \tag{8.1}$$

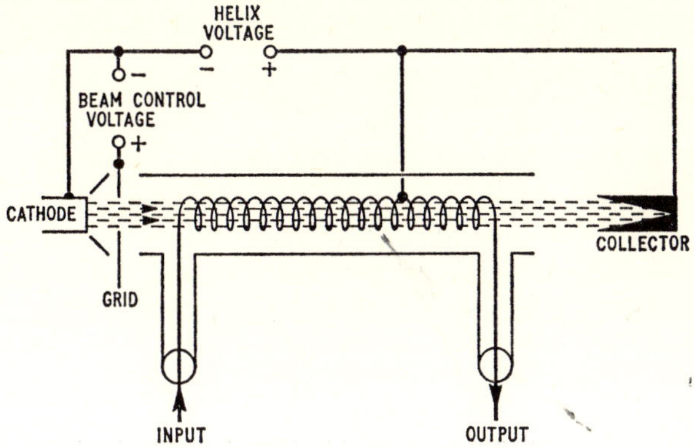

Fig. 8.1. *Helix travelling wave tube*

where ψ is the pitch angle of the helix. The r.f. electric fields which are produced are then as shown in Fig. 8.2 These are travelling fields, and provided that there are no large mismatches at the points where the r.f. power is coupled to or from the helix, the entire field pattern will travel uniformly across the diagram. This picture still

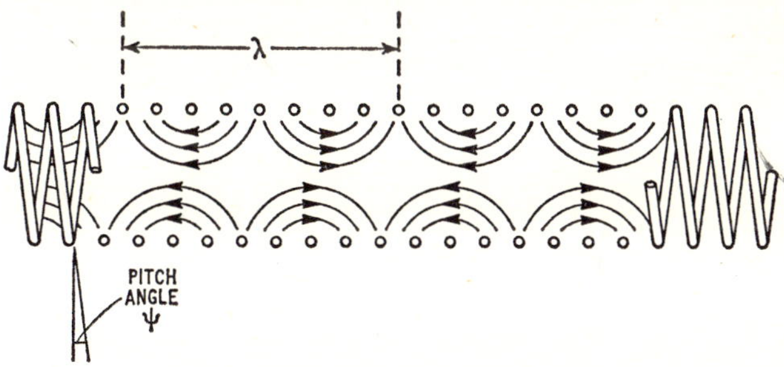

Fig. 8.2. *Electric r.f. fields inside helix*

remains true at high frequencies even if the outer conductor is not present, but at lower frequencies the velocity rapidly increases, as indicated in Fig. 8.3. In fact, in most helix tubes the helix is held inside a glass tube, which may in turn be inside a conducting tube, and the propagation is again somewhat modified by the permittivity of the glass. The form of the r.f. fields, however, is not greatly affected, and will still be as shown in Fig. 8.2.

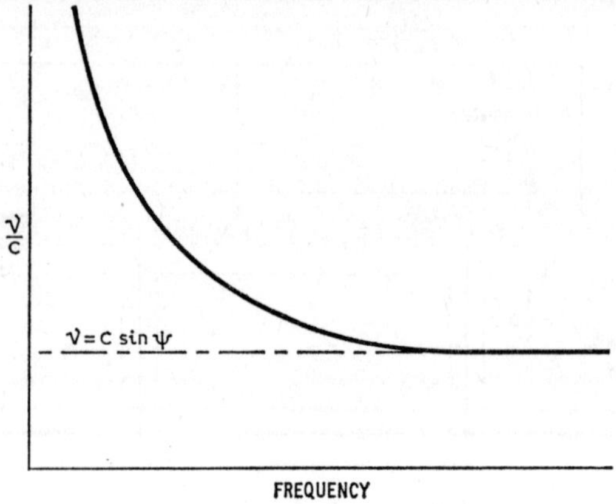

Fig. 8.3. *Propagation of a helix*

Pursuing our simple picture, let us suppose that an electron beam is injected into this helix at the same velocity as that of the r.f. wave propagating along it. To see what will happen to the beam, consider the section of it lying within one wavelength of the r.f. field, between the lines *A* and *C* in Fig. 8.4. Since the electrons and the field are travelling together at the same speed, all the electrons in the

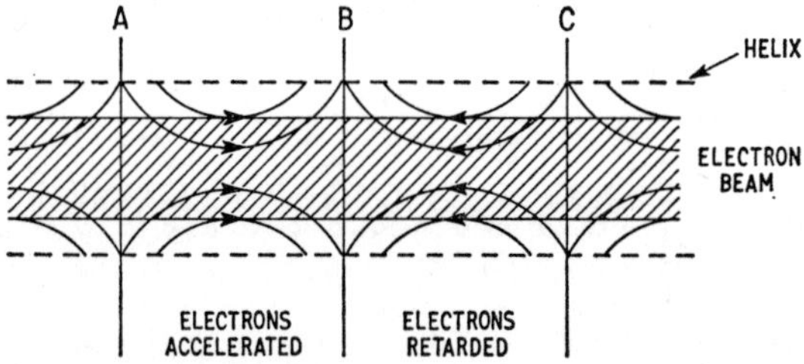

Fig. 8.4. *Bunching effect of r.f. field. Arrows indicate forces on electrons*

region *AB* will experience an accelerating force, and will gradually move towards the line *B* as they drift through the tube. Similarly all those electrons between *B* and *C* will experience a retarding force, and will drift backwards towards the centre line *B*. The net result will therefore be that all these electrons will gradually form themselves into a bunch with its centre at *B*, until the increasing space charge forces due to the higher electron density bring the process to a halt.

The next step towards the whole picture is to consider how, because of this interaction which has been taking place, the r.f. wave on the circuit has itself been changing while this bunching process has been going on. The electrons in the region *BC*, which are giving up energy to the r.f. wave by being decelerated, not only increase the amplitude of the r.f. field, but also retard its phase, a process which is assisted by the electrons in the region *AB*, even though these latter are first absorbing energy from the r.f. field. The result is that the point *B* moves steadily backwards relative to the beam, and so more and more electrons find themselves in the region of the decelerating r.f. field, and hence the interaction mechanism grows on itself: the more electrons which give up energy to the field, the greater the field becomes and the more it is retarded. Processes like this which grow upon themselves usually lead to an exponential increase with time or distance, and this is the case with the travelling wave tube. It is found that the signal increases exponentially as it progresses along the helix, until finally the process is brought to a halt by the complete disintegration of the bunch.

8.2 Coupled Waves

This is a simple ballistic picture of what is taking place. Considerably more insight into the process can be gained by consideration of the mechanism as a coupling between the various waves which can be propagated by the system, and although it will not be followed through here, this is the approach which, when made in mathematical terms, enables the performance of a t.w.t. to be predicted in great detail, at least for small signals.

It will be recalled from Chapter 3 that propagation of waves of angular frequency ω is described by

$$A = A_0 e^{j(\omega t - \beta z)} \tag{8.2}$$

where A is the amplitude of the quantity, current, voltage, field, etc, being studied. The quantity β is the phase propagation constant, and normally gives the variation of phase with distance. When the propagation is lossless, so that waves are neither increasing nor decreasing with distance, β is completely real, and is denoted by β_0 for a transmission line. It can readily be shown, for such a line, that

$$\beta = \beta_0 = \pm\, \omega\, (L_0 C_0)^{\frac{1}{2}} \tag{8.3}$$

where L_0 = series inductance per unit length
C_0 = shunt capacitance per unit length

We have also met, in Chapter 6, the result for space-charge waves on an electron beam by itself,

$$\beta = \frac{\omega}{u_o} \pm \frac{\omega_p}{u_o} \tag{8.4}$$

or

$$(\beta - \beta_e)^2 = \beta_p{}^2$$

Equation 8.3 is derived by providing that the rate of decrease in current along the line is due to charge flowing into the shunt capacitance, and that the rate of decrease in voltage along the line is due to the e.m.f. produced across the series inductance by the current. Equation 8.4 is derived, as given in Chapter 6, by supposing that the forces acting on the electrons are due to the space charge and that currents are confined to the beam.

If the two systems are considered together, so that the electrons are also affected by the electric fields produced by the wave travelling along the line, and currents are induced in the line by the electron beam, the resulting equation for the propagation constants, which for the moment are denoted by Γ, is

$$(\Gamma - \beta_e)^2 = \beta_p{}^2 + \mathrm{j}\,\frac{2\beta_e\Gamma^2\beta_o{}^2C^3}{(\beta_o{}^2 - \Gamma^2)} \tag{8.5}$$

where C is a factor we shall return to later. Equation 8.5 will in general have complex roots. We can see immediately, however, that if $(\beta_o - \Gamma)$ is very large, i.e. if waves are propagating with very different velocities on the beam and the line, then Equation 8.5 approaches the lower form of Equation 8.4, which is as would be expected. It is only when β_o is close to Γ that interaction is occurring, and Γ becomes complex.

Equation 8.5 is a fourth-order equation giving four values of Γ, which is to be expected, since Equations 8.3 and 8.4 each give two waves. One of these is the backward-going circuit wave having $\beta = -\omega(L_oC_o)^{\frac{1}{2}}$, which would not be expected to enter into any interaction. The other three solutions may be found by computation, and are conveniently expressed in the form

$$\Gamma = \beta_e\,(1 - Cy) + \mathrm{j}\beta_eCx \tag{8.6}$$

Here, a positive value of y represents a wave going faster than the beam ($\beta < \beta_e$) and a positive value of x represents a wave increasing in amplitude with distance, and we expect to find three sets of values for x and y describing how these three waves propagate as the beam velocity is varied around the circuit velocity.

A typical result for given values of circuit loss and space charge is

PLATE 7.1: *A high power two-cavity klystron and its magnetic circuit (by courtesy of M-O Valve Co. Ltd.)*

PLATE 7.2: *A four-cavity 10,000 Mc/s klystron amplifier (by courtesy of M-O Valve Co. Ltd.)*

PLATE 7.4: *An early reflex klystron (CV35)*

PLATE 7.3:
A high-power U.h.f. T.V. klystron (by courtesy of English Electric Valve Co. Ltd.)

PLATE 7.5: *A modern reflex klystron and its coaxial cavity (by courtesy of E.M.I. Electronics Ltd.)*

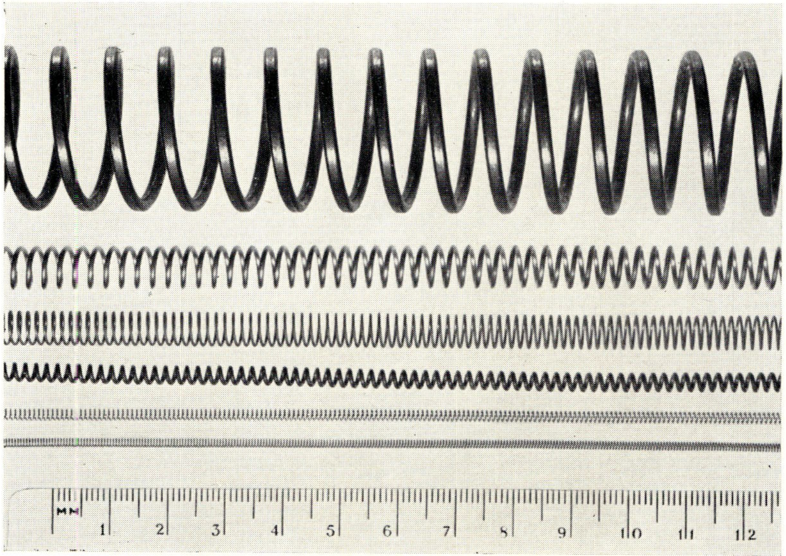

PLATE 8.1: *Travelling wave tube helices*

(*a*) *500 Mc/s 2·5 kV*

(*b*) *1,500 Mc/s 2·5 kV*

(*c*) *2,500 Mc/s 2·0 kV*

(*d*) *9,000 Mc/s 8·0 kV*

(*e*) *7,000 Mc/s 3·3 kV*

(*f*) *9,000 Mc/s 3·5 kV*

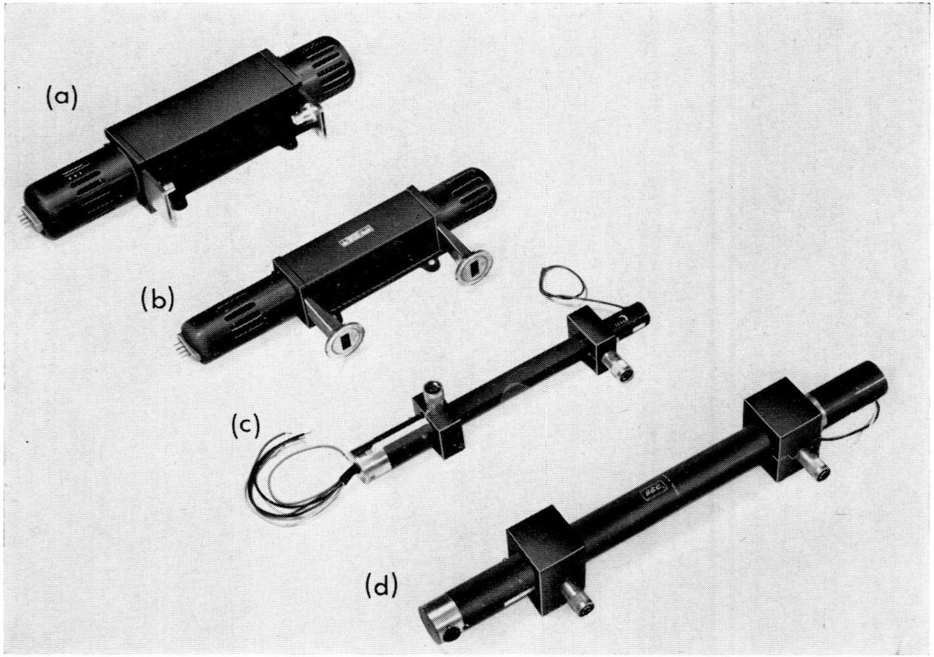

PLATE 8.2: *P.P.M. focused travelling wave tubes (by courtesy of M-O Valve Co. Ltd.)*

(*a*)	*VX 3325*	*1 W*	*4–7 Gc/s*
(*b*)	*TWX8*	*0·5 W*	*7–11·5 Gc/s*
(*c*)	*TWS9*	*250 mW*	*1·7–2·3 Gc/s*
(*d*)	*TWS10*	*20 W*	*1·7–3·0 Gc/s*

represented by Fig. 8.5(c). In this diagram the values of y are shown by the full line and the values of x by the broken line. The horizontal axis shows the variation of beam velocity (u_o) from below to above that of the circuit wave (v). On the left-hand side of the diagram there are three lines, which are similar to those shown in Figs. 8.5 (a) and (b), which show respectively the forward circuit wave and the two space-charge waves without interaction.

Looking again at Equation 8.6, outside the region of interaction $x=0$, so that

$$\Gamma = \beta_e (1 - Cy) \tag{8.7}$$

hence for the circuit wave, where $\Gamma = \beta_o$,

$$\beta_o = \beta_e (1 - Cy)$$

i.e.

$$y = \frac{1}{C}\left(1 - \beta_o \frac{u_o}{\omega}\right) \tag{8.8}$$

and this is represented by the sloping line of Fig. 8.5(a). Similarly the beam waves are given by

$$\Gamma = \beta_e \pm \beta_p \tag{8.9}$$

or

$$\beta_e (1 - Cy) = \beta_e \pm \beta_p$$

i.e.

$$y = \pm \frac{\beta_p}{C\beta_e} = \pm \frac{1}{C}\frac{\omega_p}{\omega} \tag{8.10}$$

Over this small range of u_o, ω_p can be taken as constant, so these two waves are represented by two equal and opposite constant values of y as in Fig. 8.5(b).

In Fig. 8.5(c) it is seen how these waves become modified as the beam velocity is swept through the circuit velocity. Between the two vertical lines the circuit wave remains faster (y positive) and gradually turns into the fast space-charge wave. The two space-charge waves (2 and 3) become two waves travelling at the same velocity, shown by the short sloping line, which is slower than that of the electron beam (since $y<0$). Also over this region there exist two values of x, as indicated by the top and bottom halves of the dotted ellipse, which are associated with these two equal values of y. This means that there are two waves, both travelling with the same velocity which is slower than the beam velocity, one of which is growing exponentially with distance (x positive) and the other decaying. On the right-hand side of the diagram, as u_o is increased still further, the waves now appear as the sum of Figs. 8.5(a) and 8.5(b).

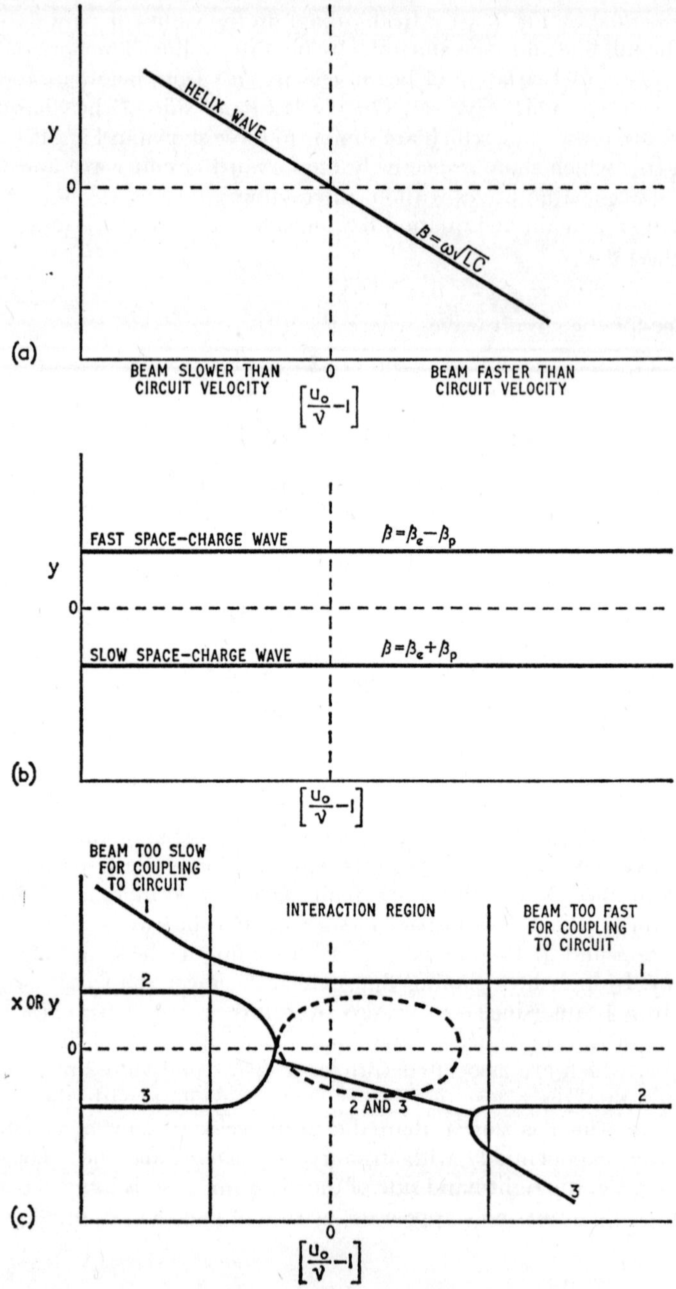

Fig. 8.5. *Beam-helix interaction*

There are thus three waves propagating in the coupled system:

(1) A wave going faster than the electron beam, neither growing nor decaying with distance.

(2) A wave going slower than the beam, whose amplitude decays with distance as described by $e^{-\beta_e Cxz}$.

(3) A wave at the same velocity as (2), with amplitude increasing with distance as $e^{\beta_e Cxz}$.

The relative amplitudes with which these waves are excited depend on several things, principally the ratio of v to u_0, but no matter in what ratio they may be excited, after progressing a relatively short distance along the helix the growing wave predominates and the other two soon become relatively negligible. From Fig. 8.5(c) it can be seen that the rate of growth of this wave (i.e. the positive value of x) starts from zero as the lowest interaction velocity is approached, rises through a maximum at some optimum voltage, and falls again to zero as the range of interaction is passed through. The effect of this on valve performance as the voltage is varied is as shown in Fig. 8.6. The t.w.t. initially has a high loss, and looks like

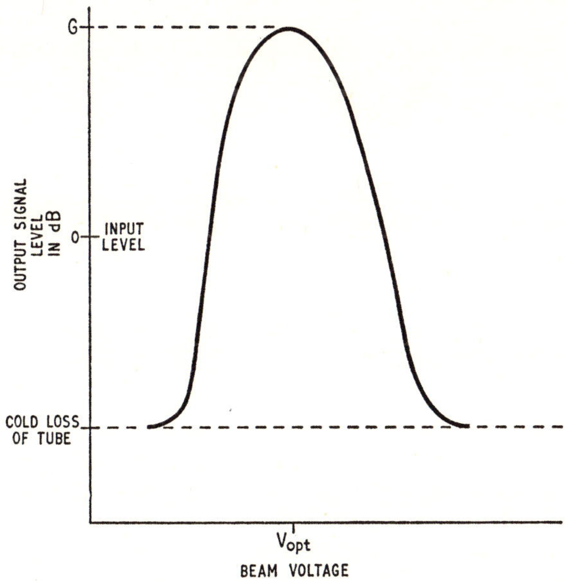

Fig. 8.6. *Variation of gain with beam voltage*

an attenuator, then as the voltage is increased the gain mechanism gradually comes into play, and the output increases, passes through a maximum at which the output may be many thousand times as much as the input, and then falls with increasing rapidity back to its original condition. The actual values of the growth constants and the optimum velocity depend on other things such as the cold loss of the valve, the amount of space charge present, the ratio of helix to beam diameter, etc, in rather complicated ways, and for any actual case the values required must be obtained by computation.

The constant C in Equations 8.5 and 8.6 is obviously an important one, since it appears in the exponent of $e^{\beta_e Cxz}$. This implies that the growing wave increases at a rate which can be expressed as BC decibels per electronic wavelength, where B is a constant which varies between about 35 and a maximum value of 47·3 for most valves, depending on the extent to which potential variations due to space charge affect the interaction. C is often called the growth parameter, and is defined by

$$C^3 = \frac{KI_o}{4V_o} \tag{8.11}$$

where

$$K = \frac{E_z^2}{2\beta_o^2 P} \tag{8.12}$$

and I_o and V_o are the beam current and beam voltage.

K is a measure of the usefulness of the circuit for this purpose, giving the ratio of the peak value, E_z, of the alternating axial electric field affecting the electrons to the power P flowing along the line. It is dimensionally equal to an impedance and is usually called the circuit interaction impedance. The way in which the total power increases along the helix is typically as shown in Fig. 8.7, where it can be seen that, after a relatively short number of wavelengths, given, in fact, approximately by $N=\frac{1}{4C}$ the growing wave is bigger than both the other two waves, and from then on is virtually the only wave present. The gain of a travelling wave tube in decibels is therefore given by $G=A+BCN$, where A is an initial loss factor, usually about 9 decibels, which takes account of the fact that the true exponential increase does not start immediately and N is the number of electronic wavelengths in the tube. At first sight therefore it looks as though the gain could be made as large as one would like. In practice a limit is set to this by the instability which arises.

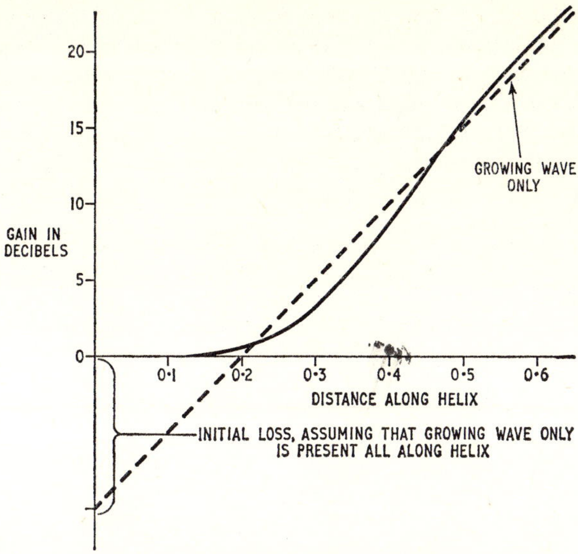

Fig. 8.7. *Actual power variation along helix*

8.3 Effect of Coupler Matching on Stability

It will be appreciated that this gain mechanism is inherently a very broad-banded one. The velocity of the wave on the helix changes only very slowly with the frequency, and even this change is to some extent compensated for by the fact that as the frequency increases the number of wavelengths in the valve also increases, so that the gain, as given by the expression above, only falls off very slowly. This is an admirable state of affairs as long as we only consider what is going on inside the helix, but, in fact, the helix is inside a vacuum system, and the r.f. power must somehow be coupled to and from it through some vacuum-tight coupling. To make vacuum-tight couplings which do not give any reflections over an almost unlimited bandwidth is not possible, and so the reflections from the couplings immediately become very important in affecting the stability of the valve.

8.4 Necessity for Attenuation

If, for example, a valve has an overall power gain A, and an amount of power P is applied to its input, then PAR_1 would be reflected from the output, where R_1 is the power reflection coefficient at the output coupling. The helix is a relatively lossless circuit, and so this power will travel back through the valve without any interaction, since it is going in the opposite direction to the beam. Similarly PAR_1R_2 will then be reflected from the input coupling and will add to the input power. The condition for oscillation in any oscillator is that the loop gain shall be greater than unity, so that if this valve is not to oscillate, AR_1R_2 must be less than 1. A typical value that might be required for A is about 10^4 (i.e. 40 dB), and if we suppose that it is equally difficult to minimise R_1 and R_2, then neither of these must exceed one-hundredth. This implies that the voltage standing wave ratio of the coupling must be less than 1·2 over the whole frequency range of the gain. Generally, this is an impossible requirement, and so the problem must be solved in some other way. The solution is to include in the helix an attenuating section, the attenuation of which is usually considerably more than the overall gain of the valve, and then to lengthen the valve to restore the overall gain to the required figure. At first thought it may appear that no advantage has been gained. However, suppose, for example, an overall gain of 40 dB is provided by an electronic gain of 90 dB plus an attenuation of 50 dB: then even if there is a short circuit at the output so that all the power is reflected along the helix, the amount arriving back at the input will be 10 dB below the incident signal, and the valve will be stable. In practice a more favourable situation exists, and the attenuation suffered by the forward-growing wave is considerably less than the 'cold attenuation' suffered by the passively propagating backward-going wave. This is because the growing wave is not propagating only on the helix, but on the combination of the helix plus the beam. If the helix were suddenly removed, the fraction of the total energy associated with the beam would continue to propagate in the form of the usual two space-charge waves, and if the beam then entered a second helix the three waves of the combined system would again be excited and a new growing-wave system would again be established. It has been shown that in this simple case, provided the drift length is kept short, the overall loss due to the severed helix can be as small as 5 dB, although the cold loss due to the severance would in principle be infinite. This approach is, in fact, adopted in some of the high-power valves described in the next chapter, but for reasons of convenience is not often met with in helix tubes, which are usually of fairly low powers.

8.4.1 Practical Attenuators

Two methods of obtaining the attenuation are in common use. Where the diameter of the helix is very small, so that it is not practicable to hold it directly inside the precision glass tubing which is the most popular construction used in these valves, it is usually held in place by three or more ceramic rods as shown in Fig. 8.8. The

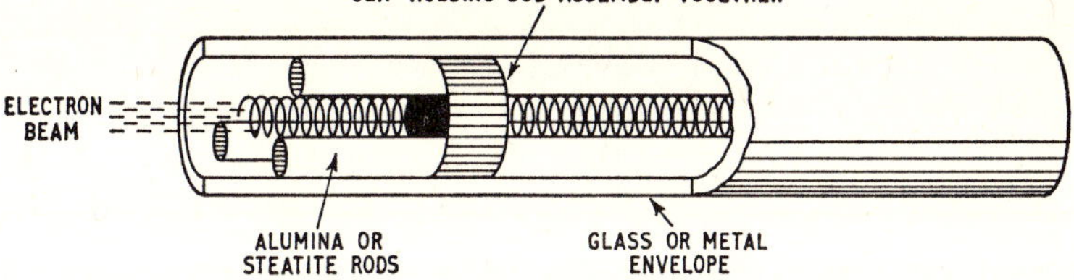

Fig. 8.8. *Typical t.w.t. construction using three rods*

attenuation is then usually obtained by coatings of 'Aquadag' or pyrolitically deposited carbon over a limited region of the rods, the coating thickness being gradually tapered off at the ends in order to preserve a good r.f. match, so that each section of the valve is itself adequately stable. Where larger-diameter helices can be used, so that the helix can lie directly inside the glass envelope, the fields of the helix can be made to couple directly to another helix wound on the outside of the glass, as in the coupled-helix coupler, shown in Fig. 8.9. This helix is then made lossy by winding it with very fine wire, and enclosing it in a resistive coating, often provided by conducting paper. This is a very convenient method, because the attenuator is one of the most difficult things to control in the t.w.t. manufacture, and at the same time it can have very critical effects on performance.

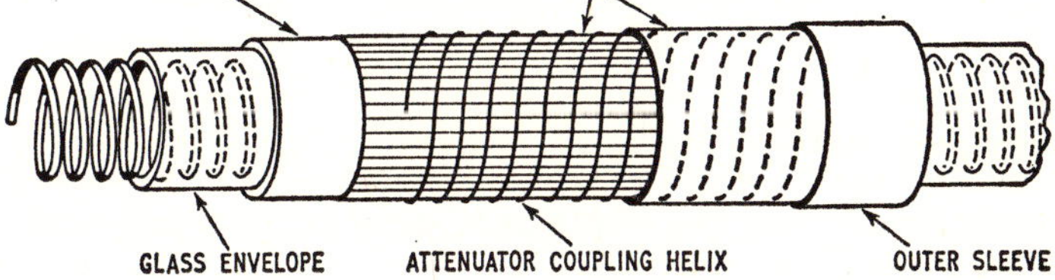

Fig. 8.9. *External coupled helix attenuator*

The critical nature of the attenuator arises in various ways. Firstly, if the actual level of the attenuation varies from one valve to another, then the gain will also vary, even though not to the same extent, and this will make it difficult for a user who desires a repeatable performance to use the valve. Then, if the attenuation starts too near the input of the valve, before the predominance of the growing wave is established, its effect on the gain becomes very large and unpredictable. An even more important effect arises if the loss extends too near the output of the valve, due to a saturation effect. The maximum power that can be obtained from a t.w.t. is, of course, limited, and in a simple valve usually arises when the efficiency is about equal to twice the value of C, the growing wave parameter. This maximum saturated power decreases very rapidly as the loss increases. If there is insufficient gain following the attenuator, then saturation will occur in the attenuator before the saturation level is reached at the valve output, and the maximum efficiency of the valve will be reduced. In order to avoid this, the gain following the attenuator must be at least of the order of 30 dB, and from the earlier discussion of stability it will be clear that the match of the attenuator now becomes of importance. Fortunately it is considerably easier to obtain well-matched attenuators than it is to match couplings, but care is still necessary to ensure that an adequate match is obtained.

8.5 Choice of Helix Dimensions

The dimensions of the helix depend chiefly on the frequency and the operating voltage, and the choice of operating voltage itself depends on the magnetic field required to focus a beam of the required power. Earlier it was seen that the gain in decibels of a given length of helix is proportional to CN. Neglecting the dispersion of the helix, the number of wavelengths N is proportional to frequency, while C for a given voltage and current varies directly with the one-third power of the coupling impedance K, which is in turn proportional to $E^2/\beta_0{}^2$. This quantity depends on the frequency. At lower frequencies, the fields extend more into the centre of a given helix than at higher frequencies. As the frequency is increased the field on the axis gradually decreases, and the energy becomes more and more concentrated closer to the helix. The variation of the effective coupling impedance at the beam will thus depend on the proportion of the inside of the helix which the beam occupies. This effect has been calculated, and the results are shown in Fig. 8.10, where CN, which is proportional to gain, is plotted against a quantity γa, which comes from the helix field analysis, for various values of b/a (the ratio of beam radius to helix radius). γ is related to β by the relation

$$\gamma = \beta\,(1 - v^2/c^2)^{3/2} \tag{8.13}$$

and so for all helix tubes, which, as shown in the next chapter, must operate below about 10 kV, γa is sensibly the same as βa. From the curves of Fig. 8.10 it can be seen that for most tubes, in which a solid beam with b/a between about 0·5 and 0·7 is used, the optimum value of βa is in the range 1·5 to 2·0. Now

$$\beta = \omega a/v \tag{8.14}$$

so for a given frequency and voltage, a is determined. What this means in practical terms can be seen from Plate 8.1, which shows some actual travelling wave tube helices over a wide range of frequency and voltage.

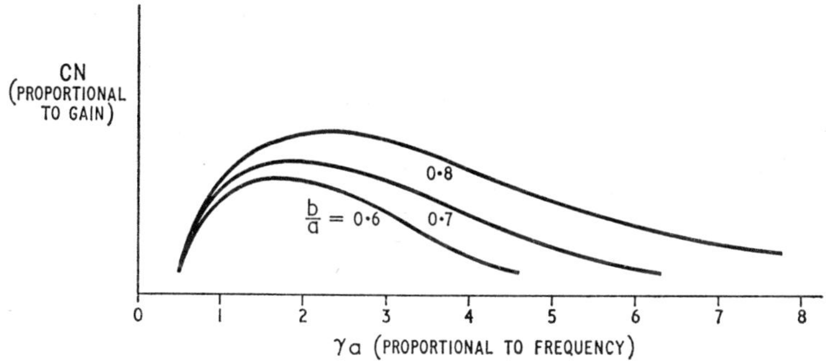

Fig. 8.10. *Variation of electronic gain with* γa

8.6 Focusing

The essential elements of the helix t.w.t. that have now been discussed are the helix and the attenuator. There is, of course, an electron beam flowing through the helix, and this must somehow be prevented from diverging and being intercepted on the helix. The subject of the generation and focusing of beams, however, is one that is relevant to all beam-type valves, and is separately treated in Chapter 10. As a general statement, it is fairly true to say that early helix-type tubes were focused by external solenoids, but that at present the great majority of new tubes, other than those where very low noise figures are required, are focused by periodic permanent-magnet circuits using permanent magnets of annular shape, made of barium-ferrite material, as shown in Fig. 8.11.

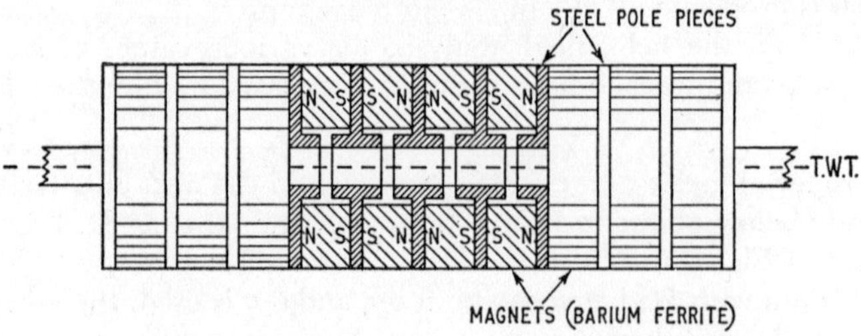

Fig. 8.11. *Periodic permanent magnet focusing system*

8.7 Collector

Two other essential elements are the collector and the r.f. couplings, which are the means by which power is fed to and from the tube.

In helix tubes, which are not usually of very high power, the design of the collector does not usually present any great problems of power dissipation. It is only necessary that it be made from a metal with high thermal conductivity, and that a fairly short path be provided to its external cooling surfaces, which are usually air-cooled vanes. What is often a more critical requirement is to avoid the emission of any fast secondary or 'reflected' electrons from it. If these are emitted, they may pass backward through the helix, and hence constitute a beam, albeit of very low current, in the reverse direction. This can provide amplification, or even just loss-free propagation of low-level signals in the reverse direction, and thus circumvent the operation of the attenuator. It is not often that this effect is large enough to give rise to gross oscillation, but it may be disastrous to the signal to noise ratio that may be required. The simplest way of overcoming this effect is by making the collector of ferromagnetic material, or enclosing it in a shield of such material, so that the magnetic field diverges in the collector. The electron flow tends to follow the direction of the magnetic field, and so the electrons strike the walls of the collector instead of impinging directly on its end surface, and by this means the effect is virtually eliminated.

8.8 R.F. Coupling

The form of the r.f. coupling depends on the frequency, and on the type of valve construction used. There are basically four forms of coupling which are in common use:

(1) A coaxial vacuum-tight seal, in which the inner conductor of the coaxial line is physically connected to the helix (Fig. 8.12).

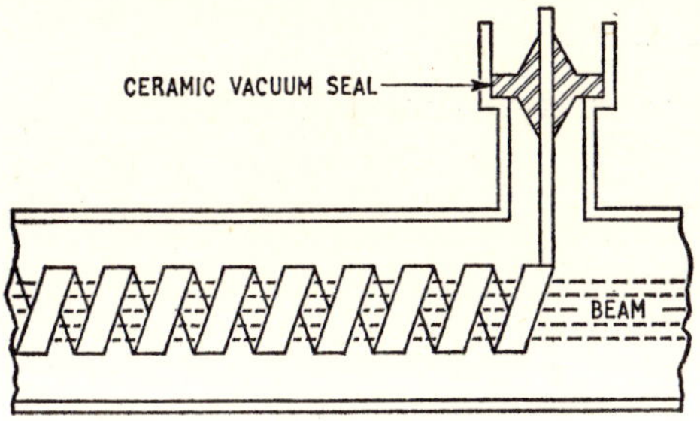

Fig. 8.12. *Coaxial direct-coupling*

(2) An external coaxial cavity, fed by a coaxial line, in which the peak fields developed across the cavity gap couple to an antenna inside the vacuum envelope, which is connected to the ends of the helix (Fig. 8.13). This type is used chiefly with glass envelope tubes, and, since it prevents the use of periodic permanent focusing, it tends to be confined to fairly low-frequency low-noise tubes.

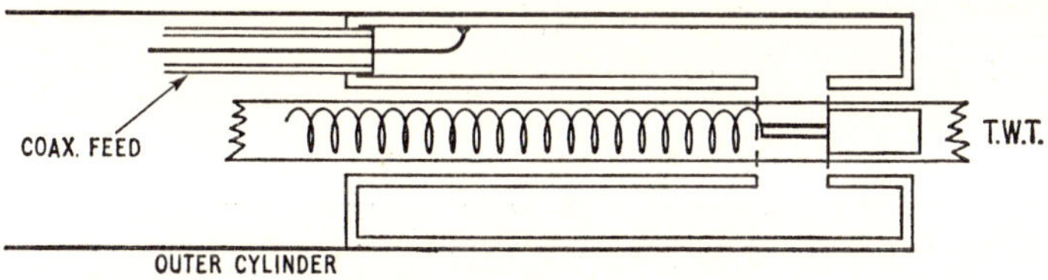

Fig. 8.13. *Coaxial cavity-coupling*

(3) A transverse waveguide, often reduced in its narrow dimension, and terminated in a short circuit beyond the coupling, which couples to an internal antenna in a manner similar to the coaxial cavity described above (Fig. 8.14). This is used in the higher-frequency ranges, e.g. above about 4,000 Mc/s, where the width of the waveguide becomes conveniently small. Because of the reduction in width, it is possible to insert the waveguide between the magnets of a periodic stack, as shown in the top two valves of Plate 8.2.

(4) An external coupled-helix coupling (Fig. 8.15). This form of coupling, which has already been referred to in regard to attenuators, depends on the properties of coupled transmission lines. If two

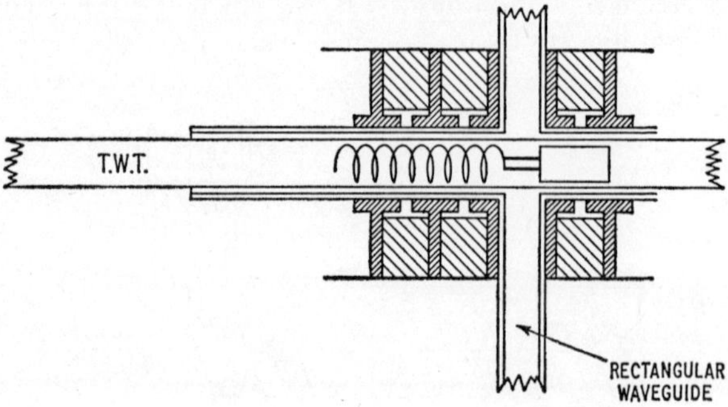

Fig. 8.14. *Waveguide coupling with p.p.m. focusing*

helices are wound concentrically, but in the opposite sense, and are dimensioned so that their propagation velocities are the same, they interact with each other in various ways, which depend on how they are excited. If they are excited with equal amplitudes and in phase, the fields between them are chiefly longitudinal. If they are excited in opposite phases, the fields are radial. These two modes propagate with different velocities. If now power is fed on to only one of the

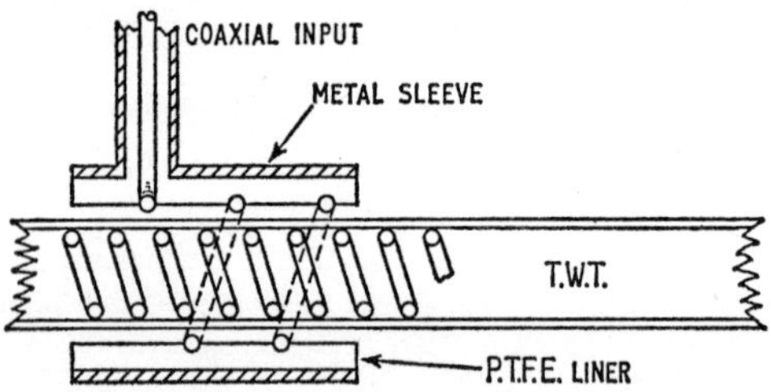

Fig. 8.15. *Coupled-helix coupler*

helices, both modes are excited, and due to their different propagation speeds, there is a periodic interference pattern set up along the length of the coupled system. In this pattern, often referred to as the *space-beat* pattern, the power is found to fluctuate alternately between the inner and the outer helix. This behaviour provides the basis of the coupled-helix coupler, in which one helix is inside the glass envelope of the tube, while the other is contra-wound on the outside of the envelope. The input power is fed to the outer helix by a coaxial line as shown and by the beating action it transfers on to the inner helix so that at one-quarter of the space-beat wavelength farther along the tube all the power is flowing on the inner helix and none is left on the outer helix. The outer helix can therefore be discontinued, and the power will then continue to flow on the inner helix. This coupler has the advantage of being directional, and if carefully designed and constructed can be placed at any point along the tube, although the inner helix may be continued beyond its ends. It is also an essentially broad-band device, since it may be expected that with proper design the relative propagation velocities of the two helices will change only very slowly. In practice bandwidths of up to ten to one in frequency have been achieved. By suitable design it can also be made to fit into a periodic focusing system, and therefore provides a very elegant solution to the coupling problem of the lower-frequency periodically focused tubes. Two travelling wave tubes using this form of coupler are shown in the lower half of Plate 8.2.

8.9 Characteristics and Uses of Helix Tubes

Helix tubes have now been made at all frequencies between 200 Mc/s and 60,000 Mc/s, and at power levels ranging from a few hundred microwatts in the case of low-noise tubes to about a kilowatt. There does not seem to be any intrinsic limit to the amount of gain that could be achieved, and it is usually the system in which the tube is to be used which defines the maximum usable gain, since amplified thermal noise from the cathode ultimately limits the signal/noise ratio as the gain is increased. In practice values of gain of up to 60dB in a single tube are encountered, and values in the range 30dB–50dB are quite normal.

Travelling wave tubes are used in various ways, in both the radar and telecommunications fields. Their general characteristics are those of a very broad-band amplifier, whose gain can be varied extremely quickly from the cold loss level of many tens of decibels to the maximum gain figure by control of the beam current. The simplest representation of performance is probably the plot of output power against input power, at a single frequency and a single value of beam current, as shown in Fig. 8.16. The upper limiting curve of this set is obtained by setting the input power level, and adjusting the helix voltage to give the maximum power output, and proceeding in this way across the entire range of input power, while the series of lower lines show the performance obtained by keeping the helix voltage fixed, and reducing the input power from that required to give the maximum output at that voltage. Since this is the way that

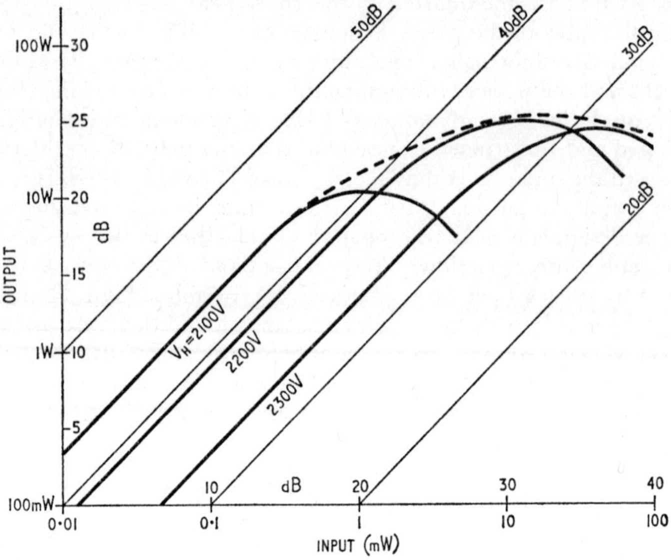

Fig. 8.16. *Input-output curves of t.w.t. at constant beam current*

most users would require to operate the valve, these lines represent operating curves. Two points may be noted: because the axes are logarithmic, i.e. are plotted in decibels above some fixed level, lines at 45° represent lines of constant gain. It can be seen that operation at constant voltage gives a region of nearly constant gain over a wide range of input power, deviating from this at about 5 dB below the maximum power output. Secondly, it may be noted that in order to obtain the maximum power it is necessary to increase the helix voltage, and hence the beam velocity, above the small-signal level, and that this, while giving an increase in power, reduces the available gain. This effect arises from the fact that as the electron bunches give up energy to the growing wave they must slow down, and so the synchronism between the wave and the beam which is required for the most efficient interaction is impaired. This can be remedied by increasing the helix voltage, so that as the electrons slow down the efficiency of interaction is at first improved and a greater transfer of power becomes possible. This improves things at the output end of the tube, giving an increase in the maximum power, but reduces the efficiency of interaction throughout the rest of the tube, thus giving the gain deterioration that is shown. In principle it is possible to prevent this situation by decreasing the wave velocity by variation of the helix pitch towards the output end. This technique is at present being applied to higher-power tubes, and has produced efficiencies as high as 50%. The value of $2C$ mentioned earlier generally gives maximum efficiency in the range 10–25% unless means such as this, or collector depression, are used to give an improvement.

Similar sets of curves can be obtained at other frequencies across the useful bandwidth, and from these a further set of curves can be constructed showing the variation with frequency of gain or power at various helix voltages. If these are studied in conjunction with another set showing the variation of gain and power at fixed voltages, but with varying beam current, an idea of the complete performance, at least regarding amplitude, can be built up. Clearly the complete measurement and specification of even one t.w.t. is a very extensive and therefore also expensive undertaking, which is, in fact, very rarely done outside research or development laboratories.

The information described above is sufficient to decide whether a t.w.t. is adequate for some noncritical applications, e.g. the use of a tube in an amplifying chain for a radar or beacon system, or for a low-capacity telecommunication system. Other uses, particularly for large-capacity telecommunication systems, are far more exacting, and for these, which actually represent by far the greatest present use of helix t.w.t.'s, the coupling matches, the noise performance, the detailed gain flatness with frequency and the phase variation with signal amplitude all become of great importance, in addition to other obvious user considerations such as ease of replacement, cooling and especially life.

Travelling wave tubes are widely used in microwave link systems as output amplifiers. Typically one r.f. channel using frequency modulation may carry up to 2,000 separate telephone channels. It is the need to keep these channels separate, i.e. to prevent cross-talk arising between them over a complete system which may contain over a hundred repeater stations, that imposes the stringent performance requirements referred to above. Coupling matches and gain flatness are interrelated. They are achieved by extreme attention to precision in pitch uniformity, attenuator distribution and circuit dimensions, and an example of what can be repeatably achieved in these respects is shown in Fig. 8.17, which refers to the 2,000 Mc/s tube type shown in Plate 8.2.

Phase variation with amplitude is a characteristic which, like the amplitude non-linearity apparent in Fig. 8.16, increases steadily as the signal level is increased, due to the steadily increasing difference

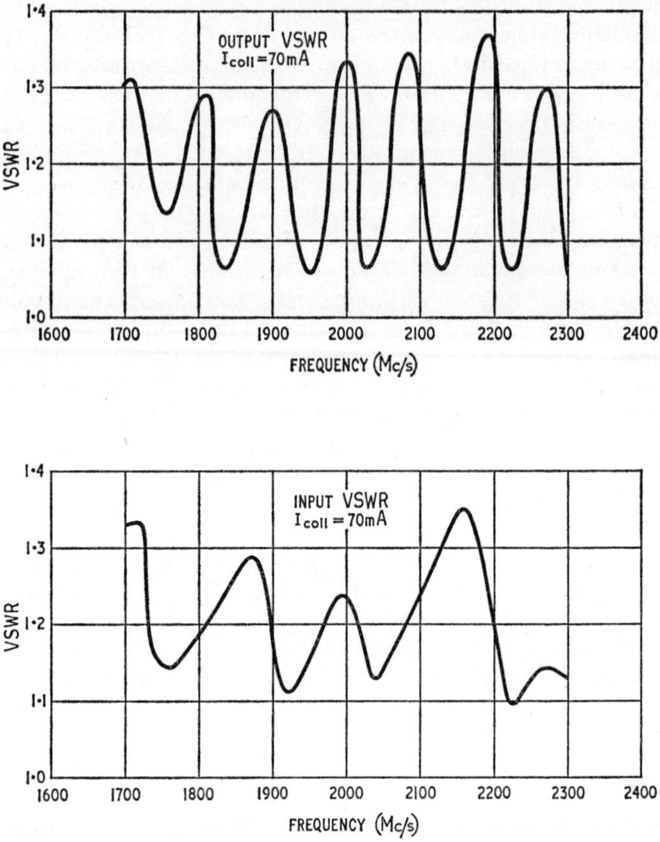

Fig. 8.17. *Hot matches of 2,000* Mc/s, *20* W *t.w.t.*

between the beam input and output velocities. The measurement of phase variation, which is considerably more difficult than that of amplitude variation, because of the high degree of stability required, may be made directly by the use of an r.f. bridge, as shown in Fig. 8.18, although more complex methods in which the actual measurement is made at a modulating frequency are often used. The pattern of behaviour which is revealed is exemplified in Fig. 8.19. In this figure curves of constant amplitude-to-phase modulation conversion at the input, measured in degrees phase shift per decibel power variation, are drawn on the input-output plot described previously.

Each of the lower curves shows the way in which the a.m. to p.m. conversion varies at a fixed helix voltage as the input power is varied, and shows the condition corresponding to the appropriate input-output curve at the same helix voltage. The required operating

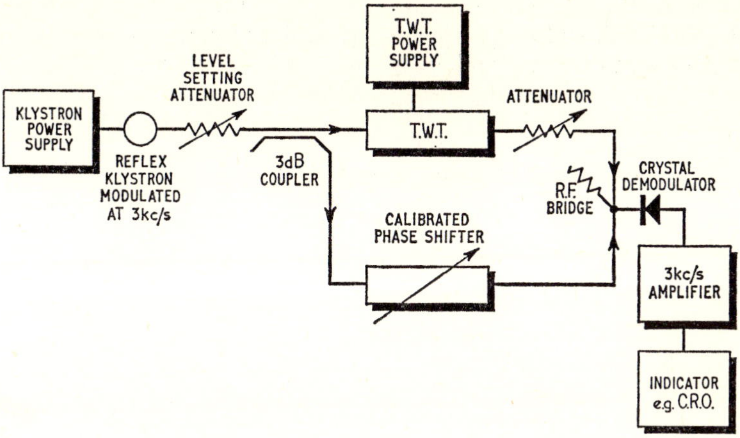

Fig. 8.18. *Circuit for measurement of phase characteristic*

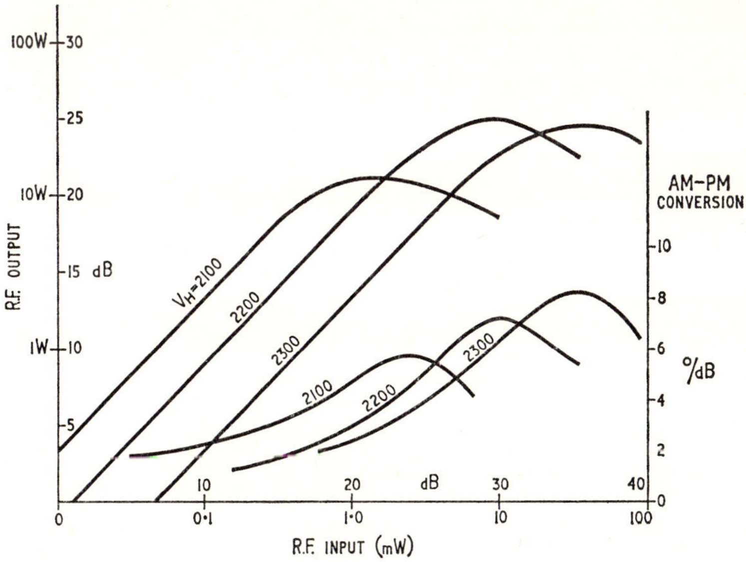

Fig. 8.19. *A.M. to P.M. curves on transfer diagram*

point is then selected to give the best compromise between the various performance characteristics desired.

Before leaving this type of valve, perhaps we may note the most recent and spectacular use in communications satellites, in which, as in the microwave repeater, the t.w.t. is used as an output amplifier.

Various other applications, such as for r.f. mixing, switching, level control, will suggest themselves to the reader, and there seems little doubt that many others will appear in the future.

BIBLIOGRAPHY

PIERCE, J. R., *Travelling Wave Tubes*, Van Nostrand (1950).

Chapter 9

HIGH POWER TRAVELLING WAVE TUBES AND OTHER BEAM TUBES USING TRAVELLING WAVE INTERACTION

9.1 Use of ω-β Diagram

The qualities required of a circuit for use in a travelling wave device are that a r.f. wave on it should produce field components in such a direction that they will interact with an electron beam and have a phase velocity over the required frequency range that is similar to that of the beam.

The representation of these qualities can be made by two curves, one giving the dispersion, or change of velocity with frequency, and the other the change of coupling impedance with frequency. The dispersion can be shown in various ways, of which the plot most frequently used is that of ω against β, as introduced in Chapter 3.

Using the ω-β diagram, as it is called, an indication of the range of frequency over which interaction with an electron beam will occur is given by adding to the diagram lines representing the properties of the beam. Since ω/β is equal to a velocity, it follows that the beam velocity itself can be represented by a diagonal line passing through the origin, as shown in Fig. 9.1. In other words, this line gives the variation of the electronic propagation constant β_e with frequency for a given beam velocity. From Equation 6.1 the fast and slow space-charge waves are then represented by the parallel dotted lines spaced to the left and right by β_p ($=\omega_p/u_o$ where $\omega_p/2\pi$ is the plasma frequency and u_o is the beam velocity).

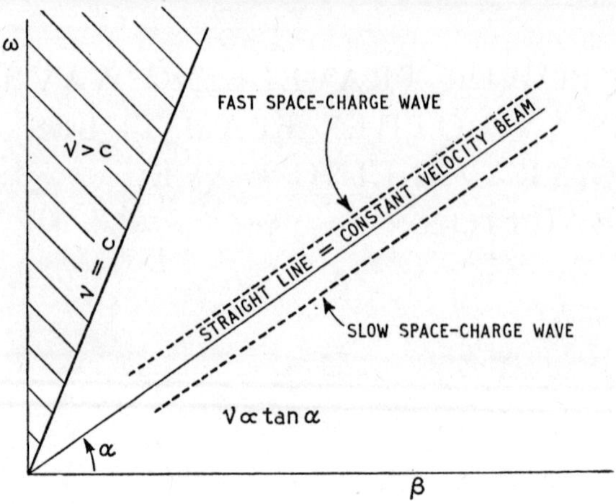

Fig. 9.1. *Representation of electron beam on ω-β diagram*

In Fig. 9.2 the curved line represents the dispersion of a typical high-power t.w.t. circuit. We know from the last chapter that the main interaction between beam and circuit will take place over the narrow frequency range between the intersections of the space-charge wave lines and the dispersion curve. Fig. 9.2 thus indicates that this circuit would provide interaction over a limited frequency range in the centre of its pass-band. If a faster beam were selected,

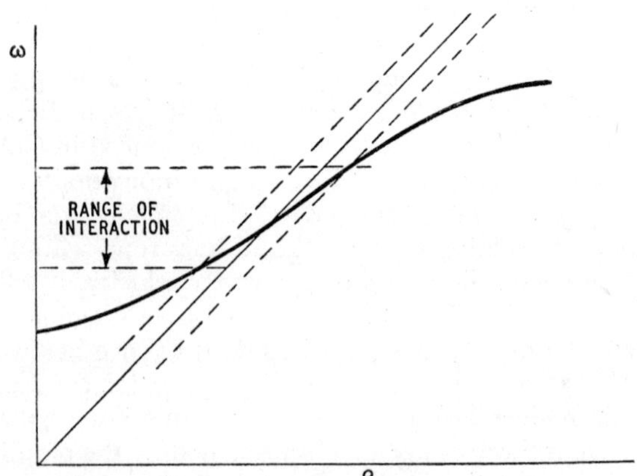

Fig. 9.2. *Use of ω-β diagram to show interaction*

which would be represented by a more steeply sloping line, the operating frequency would be decreased and the range would be reduced due to the increasing curvature of the dispersion curve. Since the maximum velocity of an electron beam would be that of light, the thick line on Fig. 9.1, which is drawn at a slope corresponding to this velocity, divides the diagram into two regions; over the left-hand region no interaction can ever take place, since the phase velocity there exceeds the velocity of light.

9.2 Interaction with Space-Harmonics

Nearly all slow-wave circuits are periodic, i.e. they are not uniform along their length like a coaxial line, but consist of a distribution of conductors so that they are similar in section only at points separated by a period p. This periodicity has some interesting and important consequences on the possibilities of electron-beam interaction with the circuit wave. These result from the fact that electrons cannot recognise phase changes greater than 2π, i.e. they are affected only by whether the field is in the direction in which they are moving or opposed to it.

If an electron is moving past a periodic structure as shown in Fig. 9.3, it will be affected by the r.f. fields only when it is in the vicinity of the spaces between the structure elements, since there can be no fields parallel to the structure at the conductor surface. When the electron or electron bunch is opposite a gap in the structure it will experience a force due to the field in the gap. When the wave is

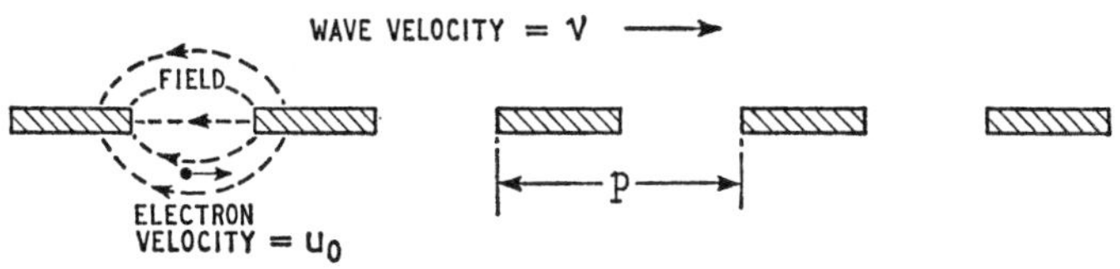

Fig. 9.3. *Space-harmonic interaction*

synchronous with the beam the electron will move to the next gap at the same velocity as the wave and will interact further with the same field. If, however, the wave is going much faster than the electron, it is possible for the field to reverse $2n$ times (where n is an integer) while the electron is shielded from it by the conducting member of the circuit, and on arrival at the next gap the field will still be in a phase favourable for interaction.

The time taken by the electron to move through one pitch p of the structure is p/u_o, and in the case considered this must be given by

$$p/u_o = p/v + 2\pi n/\omega \tag{9.1}$$

where v is the wave velocity. There are thus many velocities at which the electron beam can interact with the wave. When $n=0$ and the wave and beam are both moving at the same speed, interaction is said to be with the fundamental component of the field, while if the beam is moving slower, then at each value of n at which interaction occurs it is said to interact with the 1st, 2nd, etc, *forward space harmonic*. It is equally possible for interaction to occur if the beam and wave are going in opposite directions, corresponding to negative values of n, and interaction of this type, *backward-wave space-harmonic* interaction, is used in some of the backward-wave oscillators described later in this chapter.

An alternative approach to this concept can be made by considering the field due to the periodic circuit. If the circuit were not periodic, the field would have a sinusoidal variation only, as shown in Fig. 9.4, and would be completely described by

$$E = A(r, \theta)\ e^{j(\omega t - \beta z)} \tag{9.2}$$

where $A(r, \theta)$ describes the variation in the radial and azimuthal directions. In fact, as shown in Fig. 9.4 the field has a new periodicity imposed on it by the circuit, and as in the more familiar case of waveforms which are periodic in time, this can be dealt with by resolving it into its Fourier components. It then turns out that in order that all these components add up to give the entire field at any instant, they must be moving at different velocities, which are given by

$$p/v_n = p/v_o \pm 2\pi n/\omega \tag{9.3}$$

which is the same result as that reached by the previous argument.

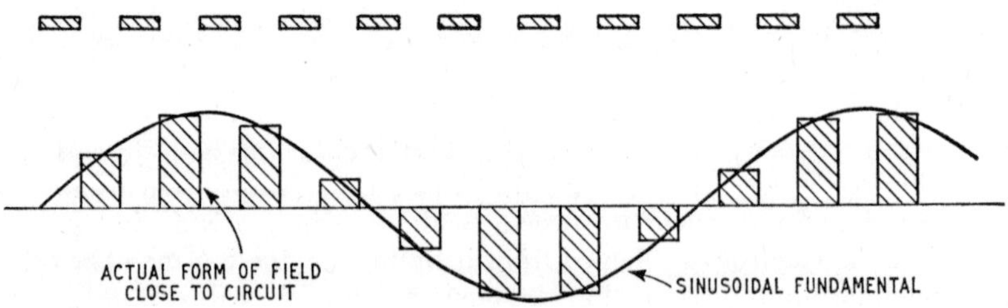

Fig. 9.4. *Periodic nature of r.f. field of periodic-circuit*

9.3 The Brillouin Diagram

These space harmonics and their interaction with electron beams can be conveniently represented by a simple extension of the ω-β diagram, due to Brillouin. Writing the velocity relation given in Equation 9.3 above in an alternative and more convenient way by multiplying throughout by ω/p, we get

$$\frac{\omega}{v_n} = \frac{\omega}{v_o} \pm \frac{2\pi n}{p}$$

i.e.

$$\beta_n = \beta_o \pm \frac{2\pi n}{p} \qquad (9.4)$$

The complete field is now described by

$$E = \sum_{-\infty}^{+\infty} A_n\ (r,\ \theta)\ e^{j(\omega t-\beta z)}\ e^{-j(2\pi nz/p)} \qquad (9.5)$$

where A_n is the amplitude of the nth space harmonic.

The fundamental propagation of the circuit can occur in both directions, and would be represented on an ω-β diagram, as in Fig. 9.5, by the curves 1 and 2 on either side of the β=0 line. In the fundamental mode the phase change per period varies between zero

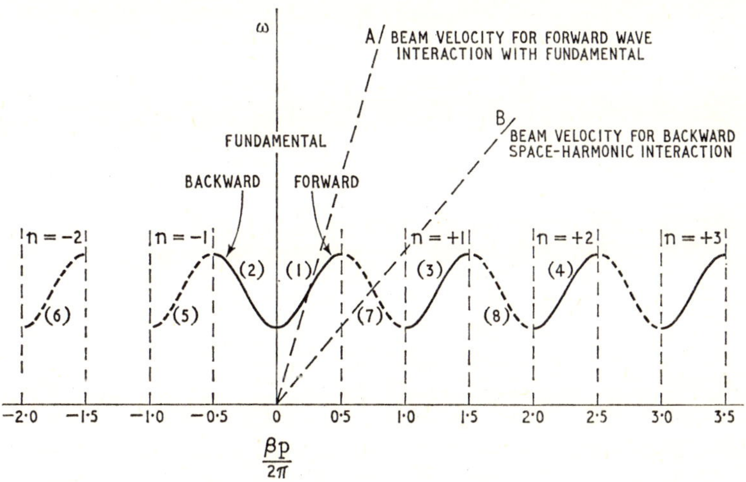

Fig. 9.5. *Brillouin dispersion-diagram*

and π, so that if the horizontal axis is scaled in units of $\beta p/2\pi$, the propagation in one direction in the fundamental mode is represented by the change from 0 to 0·5, as shown by the region 1 in Fig. 9.5. It then follows that the variation of ω with β for the space harmonics with successive positive values of n in Equation 9.5 is as shown in regions 3 and 4 in the diagram. These curves all show phase velocities in the same direction as the fundamental, and since they are all reproductions of the first region, their group velocities, as given by the slope of the curves, are all the same at a given frequency. This must obviously be so, since they are all derived from, and contribute to, the propagation of a single total wave along the structure.

Propagation in the reverse direction is represented by the region 2, in which, of course, both the group and phase velocities are negative. To interact with this reverse wave, an electron beam would also have to travel in the reverse direction, and the whole region would be exactly a mirror image of region 1 reflected in the $\beta p/2\pi=0$ axis, thus giving no further information.

9.4 Interaction with Backward Waves

The space harmonics of region 1, corresponding to successive negative values of n, will be as shown in regions 5 and 6 and will also interact with an electron beam travelling in the reverse direction. However, it is convenient to consider only the forward direction of electron flow, and these negative space harmonics are therefore mirrored in the $\beta p/2\pi=0$ axis, to give the regions 7 and 8. The particularly interesting thing about these regions is that although the phase velocities are positive the group velocities are in the opposite direction, i.e. negative.

The complete diagram on the right-hand side of Fig. 9.5 is called a Brillouin diagram. From it, the electron-beam velocities for forward- or backward-wave interaction are immediately found from the intersections with lines of constant velocity.

When the beam is interacting with a wave whose fundamental velocity is in the opposite direction to the beam, the electron bunching and the wave amplitude grow in opposite directions. The electrons which have become bunched by interaction with the wave early in their path go on to excite the circuit farther on. This process gives rise to oscillation, as described later, and the device becomes an oscillator whose frequency is determined by the electron-beam velocity. On the Brillouin diagram this is represented by a beam intersection with a downward dispersion curve, as shown in line B.

This type of dispersion curve shows *negative dispersion*, in which the wavelength is *increasing* as the frequency increases.

9.5 Propagation on a Real Helix

Interactions with backward-wave space harmonics, which, as has been seen, are present in any periodic circuit, frequently give trouble in the attainment of stable amplification from forward-wave interaction, and provide the reason why the helix, which is the simplest of all circuits, is not used at high powers, at least for pulsed operation. Its limited thermal dissipation prevents its use at mean powers above about lkW even if the best possible cooling is achieved. The dispersion curve of a helix given in the previous chapter is not, alas, really true. It arises from the analysis of a tractable model, the sheath helix, which is a tube having conductivity only in a helical direction. A real periodic helix, wound for example from metal tape, has a dispersion over the region $\beta p = 0$ to $\beta p = 2\pi$ as shown in

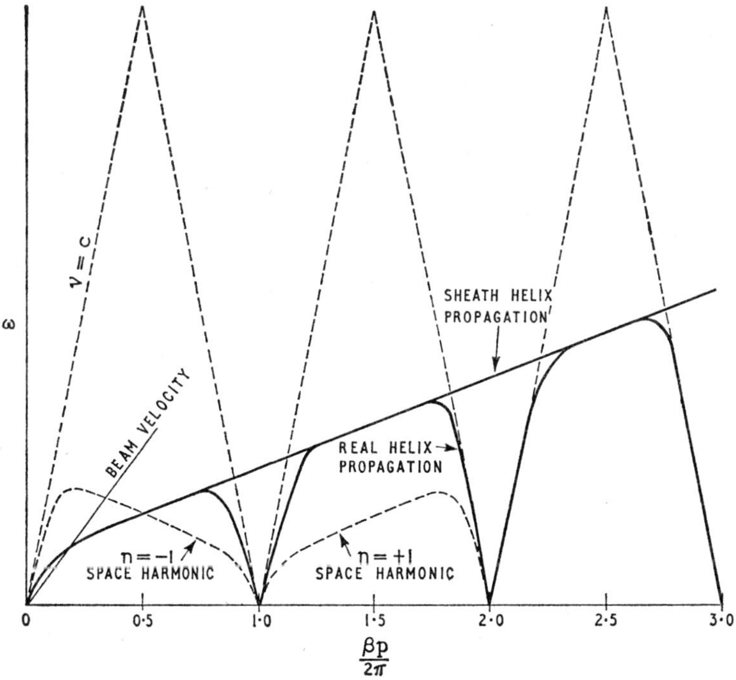

Fig. 9.6. *Propagation on a real helix*

Fig. 9.6. Its first backward space harmonic therefore will be represented by the dotted line. This means that if an electron beam is injected to work, for instance, at the point $\beta p/2\pi = 0{\cdot}2$, there would also be a possibility of backward-wave oscillation at the point where the beam line cuts the space-harmonic curve. Whether oscillation actually occurs depends on the relative amplitudes of the space-harmonic and fundamental fields, and, in fact, in this region, a stable amplifier can be made. But as $2\pi a/\lambda_0$ increases (where a is the helix radius) the relative amplitudes change, and at the point P where the two curves cross the amplitudes are equal at the helix, although due to their different distribution the fundamental is still larger on the axis. In practice, to avoid oscillation, it is necessary to design so that $2\pi a/\lambda_0$ is less than 0·5. It will be recalled that in a helix tube βa must be about 1·5 at the band centre, hence $v/\mathbf{c}$, which is nearly $p/2\pi a$, must be less than about 0·33. In practice, to obtain an adequate margin of safety, the beam voltage is made less than 10 kV. This imposes a limit on the amount of beam power that can be produced, and necessitates a large magnetic field for focusing, which may have to be supplied by a solenoid consuming much more power than the valve itself. As a result, helix tubes of greater output than 1 kW (pulsed) at 10 Gc/s, or several kilowatts at lower frequencies, are not made.

9.6 Other Two-Conductor Circuits

An attempt to overcome this situation is based on the fact that the backward-wave field has a variation around the axis, while the fundamental component has not. Two oppositely wound helices will therefore give fundamental fields which will add, while the backward-wave fields will cancel. This produced the cross-wound helix circuit, shown in Fig. 9.7, and by an engineering modification from it, the ring-and-bar circuit shown in the same figure. These circuits thus escape from the low-voltage restriction of the helix, and are used in tubes up to about 70 kV, giving pulsed powers of a few hundred kilowatts. They do not preserve the extremely wide bandwidth of the helix, and generally give bandwidths of only about 20 per cent.

Another circuit in this class, which has chiefly been used in plane form, with a strip beam, is the meander line also shown in Fig. 9.7. Indeed, the ring-and-bar circuit might more accurately be thought of as two of these joined together. This circuit provides a good example of the existence and spatial distribution of backward-wave space harmonics. At the frequency at which the phase change per gap is π, at some instants the charge distribution along the line, and

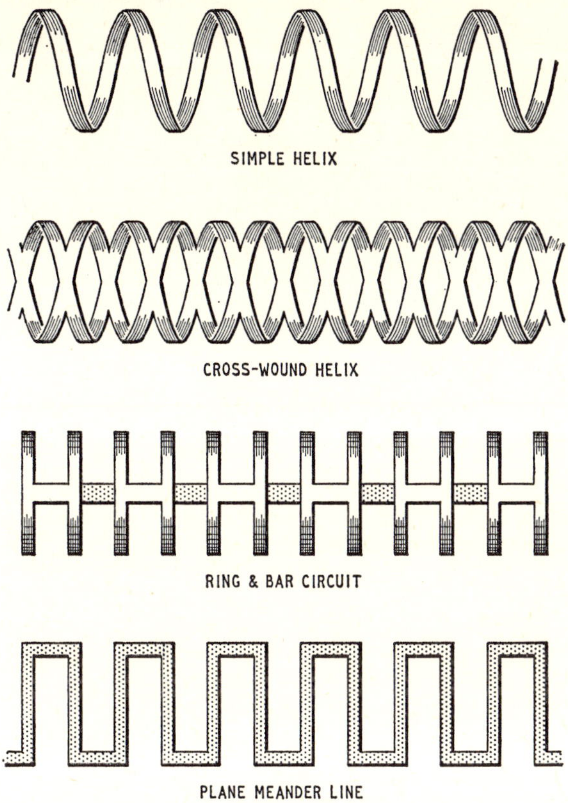

Fig. 9.7. *Some slow-wave circuits*

the field distribution, will be as shown in Fig. 9.8(a). At instants a quarter of a r.f. period later the charge pattern flowing along the line will have reached the state shown in Fig. 9.8(b). The electric field along the axis has now fallen to zero, but measured along the edge of the line there is a phase change per gap of 2π, which is indistinguishable from zero. If the frequency is now decreased, bringing the charge centres on the conducting surface farther apart (since it is roughly true that the wave travels along the strip at the velocity of light), the phase change per gap along the axis will decrease, falling to zero as the frequency becomes very low. The phase change along the edge, however, which starts from zero, can only increase, and as shown in Fig. 9.8(c) the value of βp for both

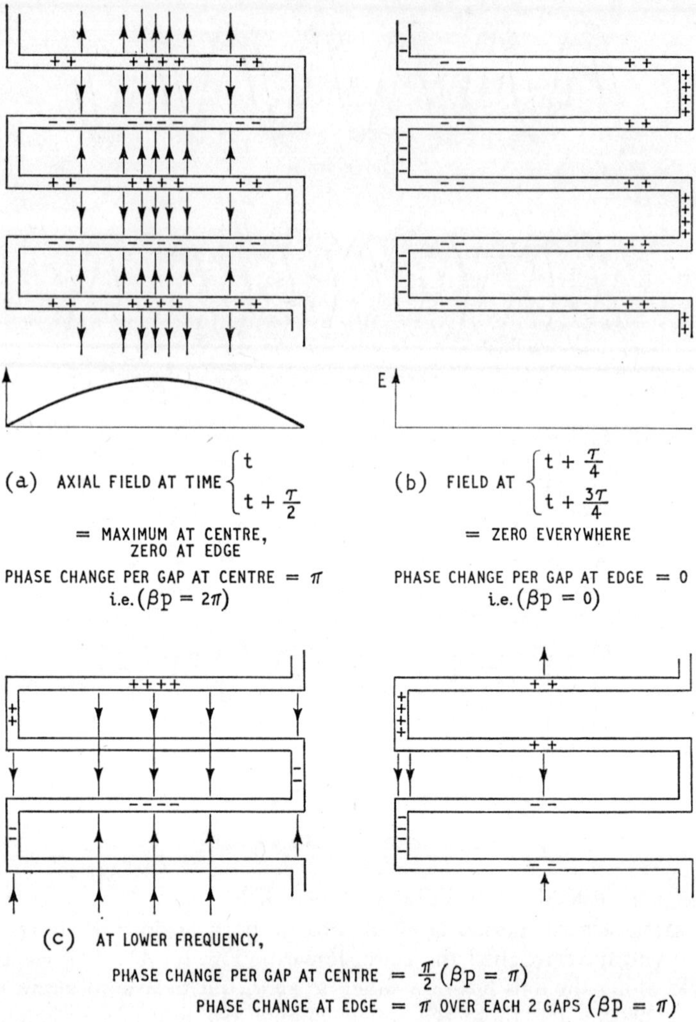

Fig. 9.8. *Fields of meander-line*

has reached π at the same frequency, although the maximum values of the two fields are still occurring at instants a quarter of a r.f. period apart in time. This behaviour is shown on the ω-β diagram of Fig. 9.9, where it can be seen that the field along the edge has a backward-wave characteristic, while that at the centre is forward

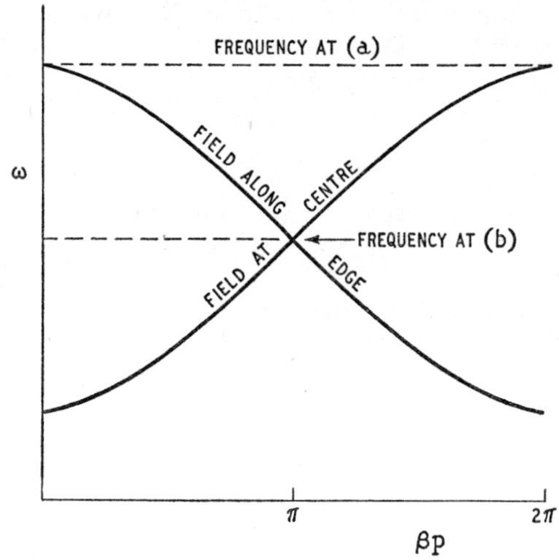

Fig. 9.9. *ω-β diagram at centre and along edge of meander-line*

wave. The implication of this is that if we wished to use this circuit as an amplifier we would send the beam along the centre, while if we desired a backward-wave oscillator the beam would be sent along the edge.

9.7 Coupled Cavity Circuits

In all the foregoing circuits there is a slow-wave structure which is contained within and insulated from a conducting cylinder or enclosure. This makes it difficult to obtain good heat dissipation from the structure, which is heated by the r.f. power it carries and also by some bombardment it receives from the electron beam. One attempts to minimise these by making the circuit of (or plating its surface with) some good conductor, and by paying great attention to the focusing of the beam. But even a very good focusing system will allow up to about one per cent of the beam to fall on the circuit, especially under maximum power conditions, when the beam consists of r.f. bunches whose charge density is much greater than that of the smooth beam which left the electron gun. Some improvement in the thermal dissipation can be obtained by making the supports of the slow-wave structure of a ceramic with high thermal conductivity, but the extent of this cannot be increased without reducing the slow-wave velocity and coupling impedance. For this reason these circuits are used particularly at the lower frequencies and mainly for pulsed amplifiers, where the mean power may be only one-thousandth of the peak power during the pulses, which are usually only a few microseconds long.

The much higher thermal dissipation required for high-power c.w. valves can only be obtained by having a continuous conductor from the inside of the circuit to the outer body, which requires the use of circuits derived from waveguides or cavities. We have already met waveguide propagation in Chapter 3, together with the use of resonant cavities. In a slow-wave structure the two to some extent join together.

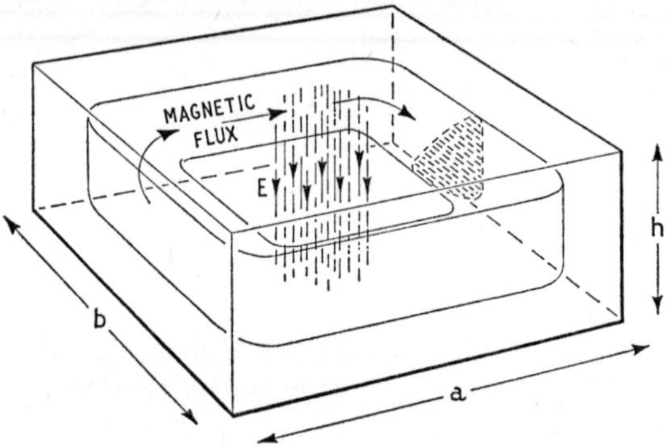

Fig. 9.10. *Rectangular resonant cavity*

A short rectangular box like that shown in Fig. 9.10 can be regarded as a short-circuited length of H_{01} waveguide, propagating in the '*a*' direction, in which case the condition for half-wave resonance is

$$\frac{1}{(2a)^2} + \frac{1}{(2b)^2} = \frac{1}{\lambda_o{}^2} \tag{9.6}$$

Clearly this is also the condition that it is a half-wavelength long if it is regarded as a waveguide propagating in the same mode in the '*b*' direction. In this mode there is a π phase change between the upper and lower surfaces, so that the box could equally be regarded as a half-wave section of guide in the '*h*' direction, but in this case a different waveguide mode is involved in which there are electric fields parallel to the direction of propagation and one half-wave variation of the field pattern in both the transverse directions. This is the *transverse magnetic* $(TM)_{11}$ or E_{11} mode. If we now make the box a square one, so that $a=b$, while h remains rather smaller, and

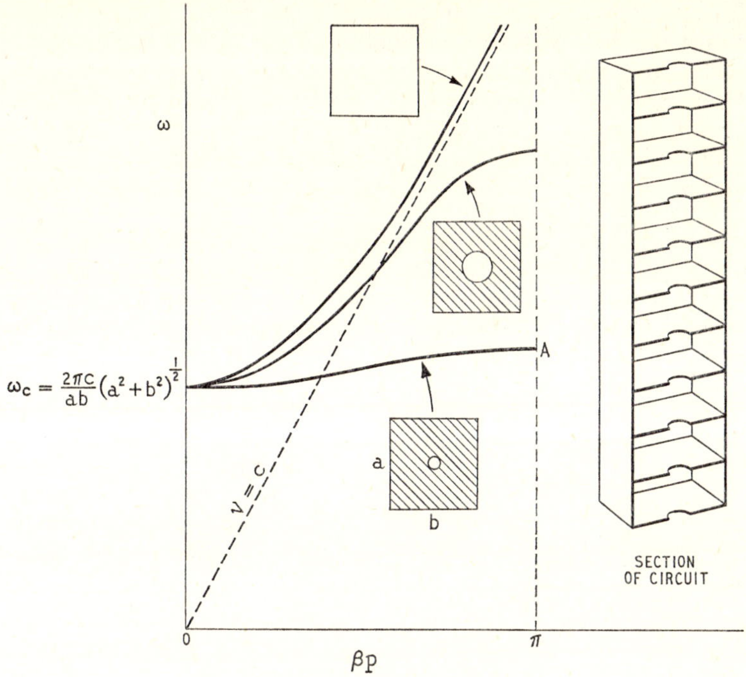

Fig. 9.11. *Dispersion of coupled-cavity circuit*

consider a series of such resonators assembled as in Fig. 9.11, coupled together by small holes in the axis and loosely fed by probes through holes in the end cavities, we should expect to find a very narrow pass-band centred on the resonant frequency of one cavity. By making the feed couplings very loose, the structure is made to behave as a short-circuited line, which will resonate whenever it is an integral number of half-wavelengths long. In this way, by taking a length of, say, ten sections, ten resonances across the pass-band can be found which represent in increasing order of frequency, phase changes per cavity of $\frac{1\,\pi}{10}, \frac{2\,\pi}{10}, \frac{3\,\pi}{10}$, etc, and from these measurements the dispersion curve shown in the lower curve of Fig. 9.11 can be plotted. The frequency about which this pass-band is centred is the cut-off frequency of E_{11} guide having the same dimensions, and from Equation 9.6, this is

$$f = \mathbf{c}\left[\frac{1}{(2b)^2} + \frac{1}{(2a)^2}\right]^{\frac{1}{2}} \tag{9.7}$$

Such a circuit could, in principle, be used for a t.w.t. by shooting a beam along its axis, but it would have only a very narrow bandwidth, and would also be very liable to oscillation at the comparatively near-by frequency of the edge of the pass-band where the group velocity is zero and, hence, the stored energy may be very high. The circuit-coupling impedance K may be written as

$$K = \frac{E^2{}_z}{2\beta^2 W_o v_g} \tag{9.8}$$

where W_o=power stored/unit length and v_g=group velocity. Hence if the edge of the pass-band, where v_g=0, occurs where E_z does not simultaneously approach zero, K can become very large, and so oscillation can occur. Such oscillations, which are often difficult to avoid in coupled-cavity tubes, are called band-edge oscillations. The pass-band of this circuit can obviously be extended by increasing the size of the coupling holes, and the effect of this is principally to raise the frequency of the π point (i.e. the point A in Fig. 9.11, where the phase change per gap is π), until ultimately, as the separating walls or irises disappear, the circuit turns into a simple E_{11} wave guide, having a phase velocity always above that of light, and hence no use for electron-beam interaction. These changes are shown by the upper two dispersion curves in Fig. 9.11. In addition, as the hole size is increased, the coupling impedance is reduced, and so less gain and efficiency would be obtained.

The two unwanted results, then, of the attempt to increase the bandwidth are that the phase velocity becomes increasingly fast, while the impedance is steadily reduced. It is therefore necessary to find other ways of increasing the bandwith.

9.7.1 The Folded Waveguide Circuit

The coupling between adjacent cavities could also be increased by putting holes in the iris wall near the edge, where the r.f. magnetic field is at a maximum, as shown in Fig. 9.12. The effect of this is to reduce the frequency of the π point, and, as the length of the slots is increased, the circuit eventually looks like a folded waveguide propagating in the fundamental H_{01} mode, with a cut-off frequency given by

$$f = \mathbf{c}/2a$$

This has increased the pass-band as desired, but unfortunately, as indicated by the dispersion curve, the circuit has now become a

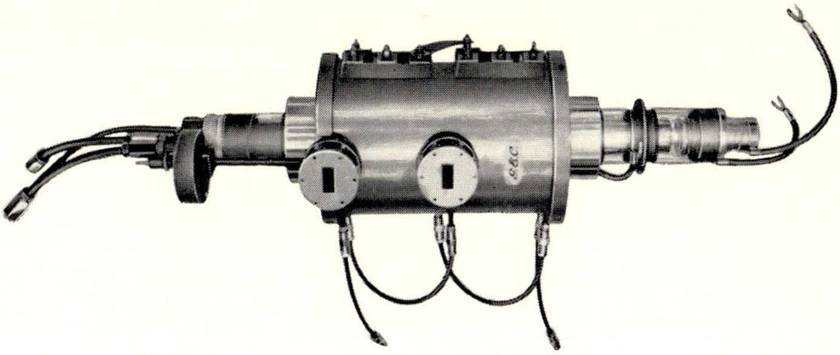

PLATE 9.1: *High-power 6,300 Mc/s travelling wave tube using cloverleaf structure (by courtesy of M-O Valve Co. Ltd.)*

PLATE 9.2: *The ophitronon electrostatically focused b.w.o. (by courtesy of M-O Valve Co. Ltd.)*

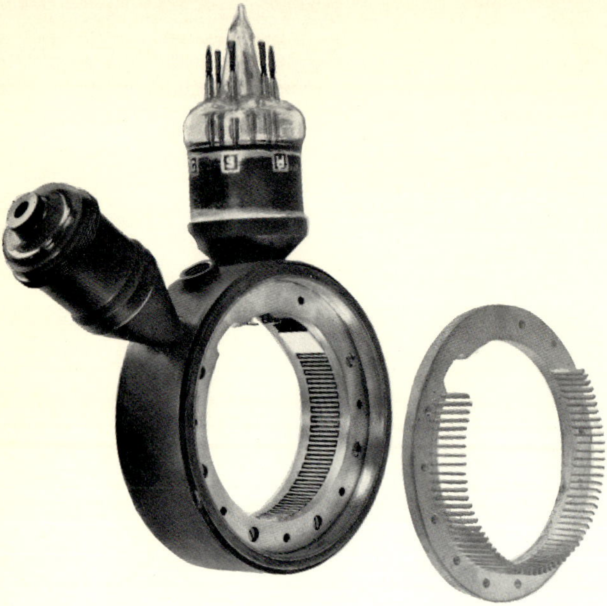

PLATE II.1: *3,000 Mc/s high-power crossed-field backward-wave oscillator (by courtesy of M-O Valve Co. Ltd.)*

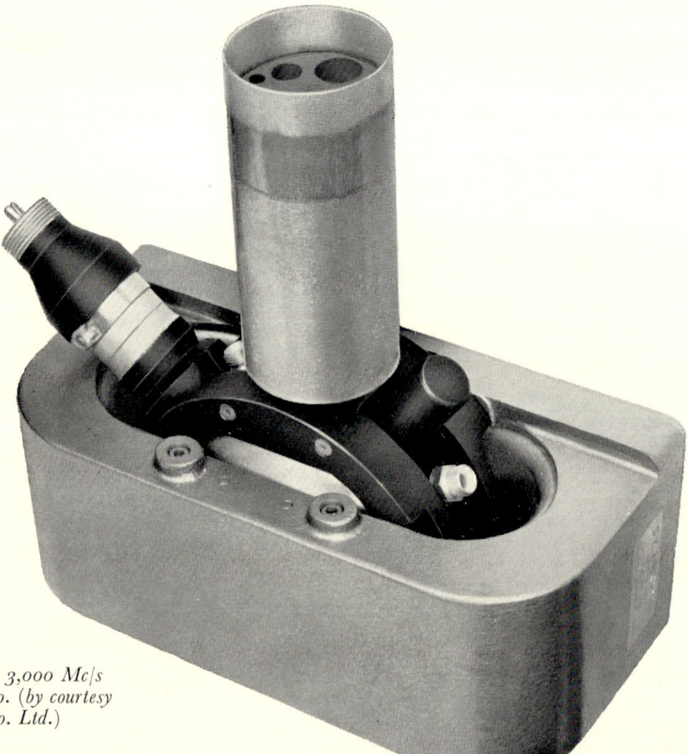

PLATE II.2: *A 3,000 Mc/s crossed-field b.w.o. (by courtesy of M-O Valve Co. Ltd.)*

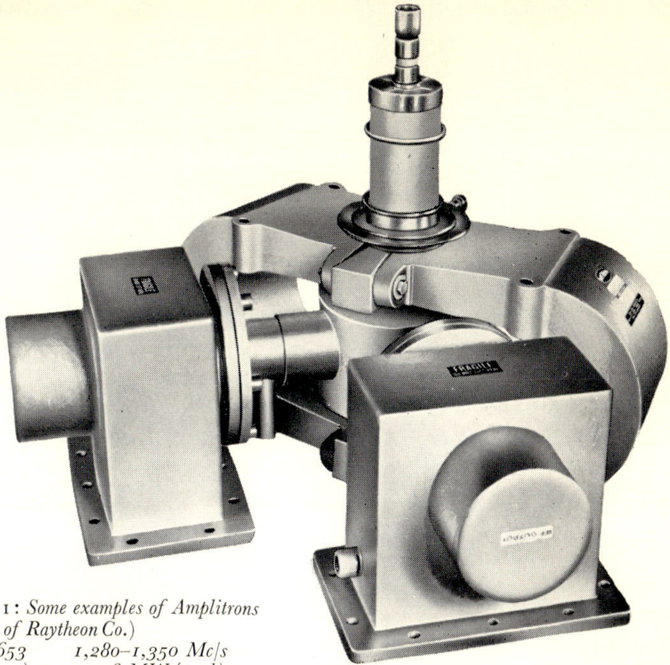

PLATE 12.1: *Some examples of Amplitrons (by courtesy of Raytheon Co.)*
(a) QKS 653 1,280–1,350 Mc/s
3·6 kW (mean) 5·26 MW (peak)

(b) QKS 622 2,900–3,100 Mc/s
15 kW (mean) 3 MW (peak)

PLATE 12.1 (contd.) (*c*) *QKS 807 5,400–5,900 Mc/s 5 kW (mean) 3 MW (peak)*

PLATE 12.2: *VX block 3168 (by courtesy of M-O Valve Co. Ltd.)*

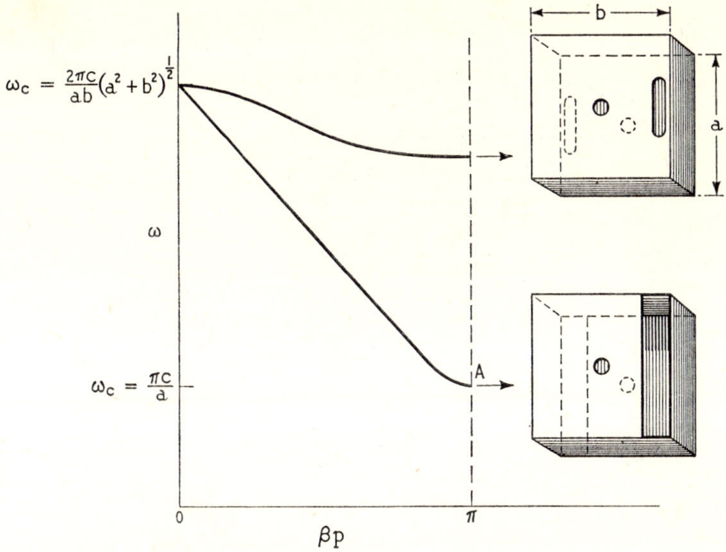

Fig. 9.12. *Coupled-cavity circuit with magnetic coupling*

backward-wave one. It may still, however, be used in an amplifier by using the first forward-wave space harmonic, as shown in Fig. 9.5.

This folded waveguide circuit, which in many respects resembles the interdigital line circuit used in backward-wave oscillators, is the first circuit that has been mentioned in which the *fundamental* propagation is a backward-wave one, and it may be enlightening to look at the reason for this.

In a waveguide at the cut-off frequency there is no phase change along the length; the phase velocity and hence the guide wavelength are both infinite. When the guide is folded into the shape of Fig. 9.13, a *geometric phase reversal* appears along the axis, and the direction of the axial field is reversed in adjacent gaps. The phase change per gap that would be seen by an electron beam is therefore π, giving a phase change per period (which includes two gaps) of 2π (Fig. 9.13(a)). As the frequency is raised, the wavelength measured along the folded guide gets shorter. At the frequency when the individual cavities are one half-wavelength long (which is recognized as the resonant frequency of the cavities by themselves) the direction of the field in alternate gaps is reversed once by the phase change along the waveguide and again by geometric reversal of the guide

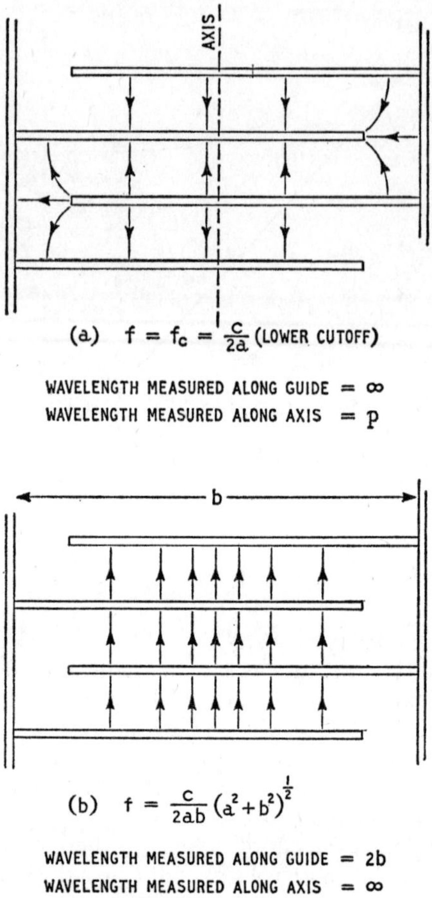

Fig. 9.13. *Folded-waveguide circuit*

direction, so that the total change per gap becomes 2π, i.e. zero (Fig. 9.13(b)). Over this frequency range therefore the axial wavelength increases as the frequency increases. If there were no reflections from the folds in the waveguide, the wavelength would then start to decrease again as the frequency was further increased, and the complete dispersion would be represented by the left-hand curve in Fig. 9.14. In fact, there is usually some reflection from the bands, and as the coupling slots are reduced in size, the stop bands due to

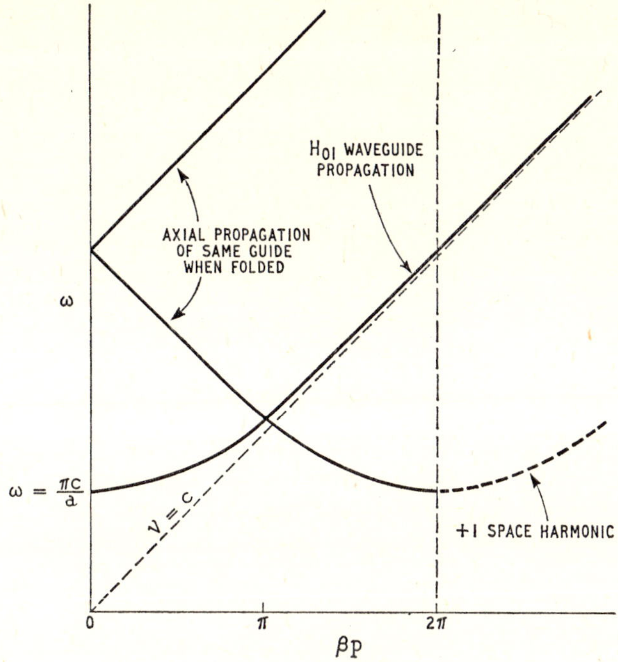

Fig. 9.14. *Propagation of folded-waveguide*

these reflections increase, until ultimately the original coupled cavity situation is again reached, as indicated in Fig. 9.11.

One result of using the first space harmonic for interaction is that the beam velocity is reduced, hence the folded waveguide circuit, usually made in a cylindrical form as shown in Fig. 9.15, is used chiefly for high-mean-power amplifiers, e.g. for c.w. powers of a few kilowatts. It is possible to use the irises forming the slow-wave structure as the pole pieces of a periodic permanent-magnet focusing system, making a very compact and convenient system.

9.7.2 The Clover-Leaf Circuit

So far we have seen that increasing the electric field coupling due to the central holes preserves the forward-wave characteristic of the coupled-cavity circuit, but makes it unsuitable by greatly increasing the velocity and reducing the impedance, while increasing the magnetic coupling by slots at the edge increases the pass-band, but

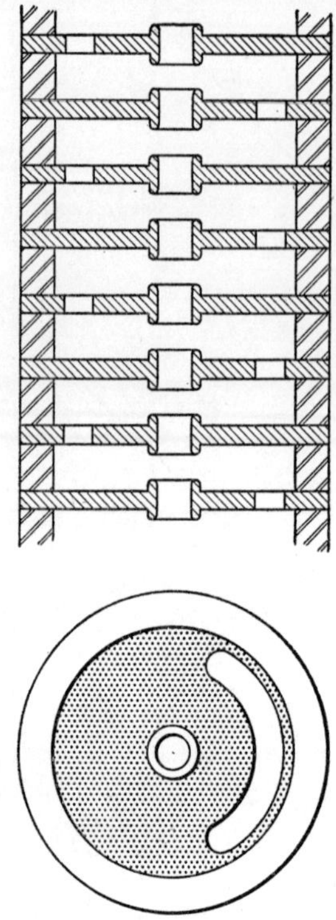

Fig. 9.15. *Cylindrical folded-waveguide circuit*

makes the circuit a backward-wave one due to the phase reversal that is introduced at the low-frequency end, which is decreased in frequency. Two ingenious ways have been devised, and are in use, in which the increased bandwidth obtained by adding magnetic coupling is obtained; the form of the coupling introduces a π phase change between cavities at the cavity resonant frequency, as occurs with the electric field coupling, which results in preserving the forward-wave characteristic. The first of these is the 'clover-leaf' circuit, shown in construction in Fig. 9.16, which indicates the shape of the individual cavities and the resulting distortion of the r.f. magnetic field. This field couples to the adjacent cavity by the radial slots in the separating iris, but in the latter cavity the lobes are rotated through 45°, and the magnetic-field components coupled through the slots add to form a field pattern in the opposite direction

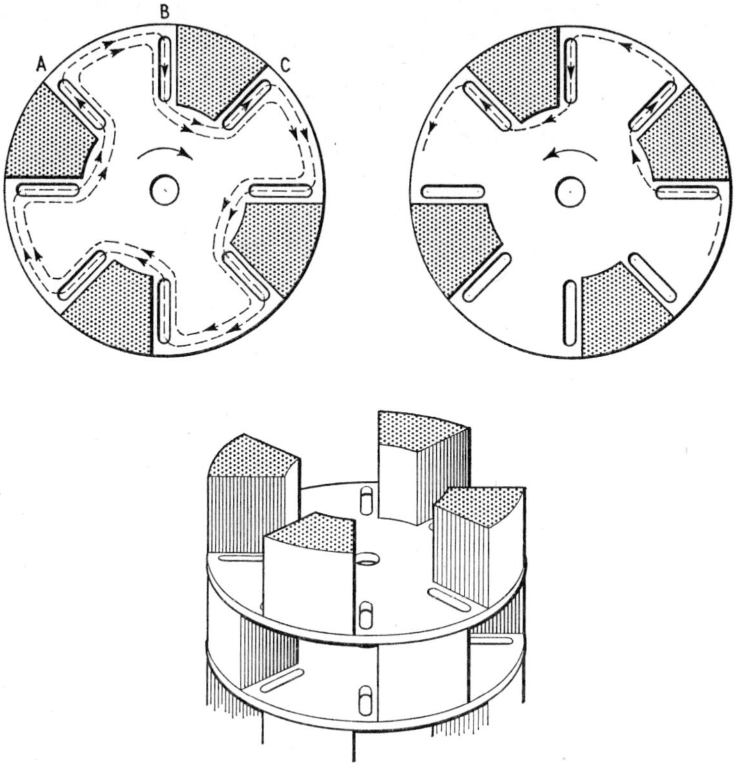

Fig. 9.16. *Clover-leaf circuit*

giving the desired phase change. This circuit has been very successfully used in the U.K. and the U.S.A. for high-power travelling wave tubes and has produced the highest powers, both c.w. and pulsed, that are available at present. A valve of this type, operating at 6,300 Mc/s with an output of 4 kW was used in the first ground satellite communication transmitter at Goonhilly Down. The corresponding transmitter in the U.S.A. used a 2 kW tube of the cylindrical folded waveguide type, using space-harmonic interaction, described in the previous section.

9.7.3 The Centipede Circuit

The second method of achieving the required phase change between cavities is by the use of coupling loops going in opposite

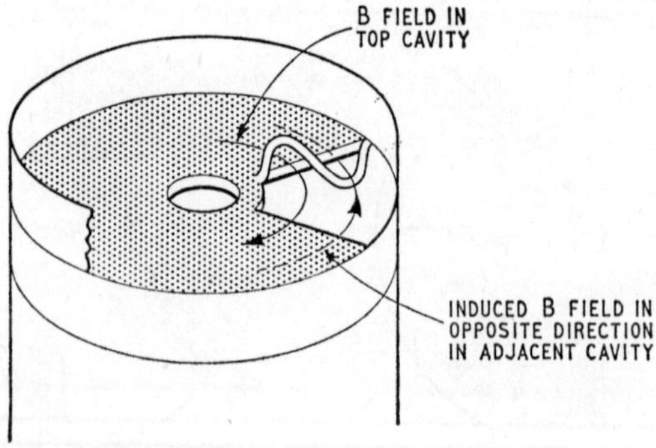

Fig. 9.17. *Coupling arrangement of centipede structure*

directions on either side of the iris, as indicated in Fig. 9.17. The r.f. magnetic field on one side then induces currents in the loop which excite a r.f. magnetic field in the opposite direction in the next cavity. As it is used, the entire wall of the iris is filled with the loops, as in the constructional diagram shown in Fig. 9.18, in order to increase the coupling to the desired level. For obvious reasons this circuit is known as the 'centipede' structure.

Unfortunately both these arrangements introduce attendant drawbacks. In the clover-leaf circuit it is possible to have a troublesome higher-order mode, having the field pattern shown in Fig. 9.19. This mode also has a large electric field in the direction of the axis, and hence it will couple to the beam, and may possibly give parasitic oscillations with output at a frequency about $1\frac{1}{2}$ times the frequency being amplified. The bandwidth also becomes rather restricted unless the voltage is kept high, about 100 kV, and so the c.w. tubes which work at a voltage of only about 30 kV are narrow-band tubes with an operating bandwidth, i.e. keeping all supply voltages fixed, of only about 2 per cent, although it is possible to tune the centre of this band over a range of about 10 per cent by adjustment of the beam voltage.

In the centipede arrangement a similar difficulty arises due to resonances of the coupling loops, which are inevitably longer than the radial dimension of the plain cavity.

9.8 Severed Circuits

Like the helix tubes, all these high-power tubes need an attenuator for stable operation if the gain exceeds about 15 dB, but owing

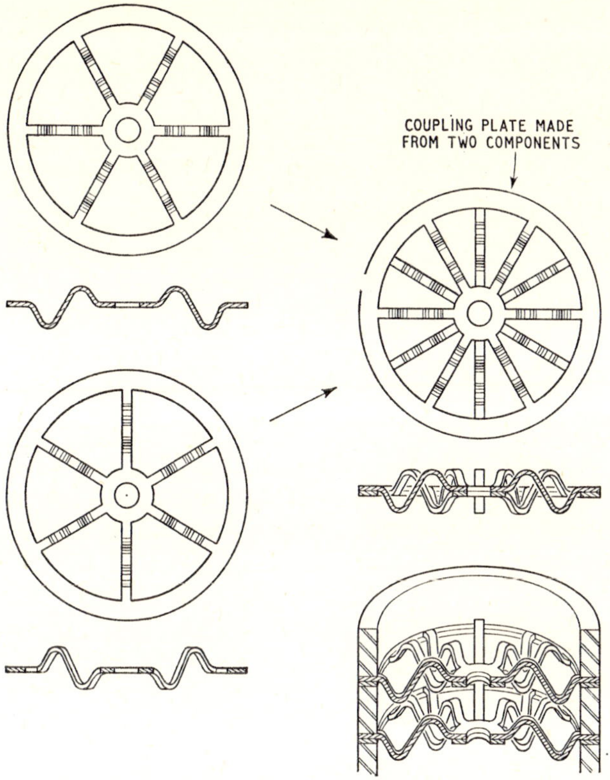

Fig. 9.18. *Construction of centipede structure*

to the amount of power available at their outputs, which may all be reflected back into the tube if the output circuit is mismatched, it may be impracticable to use an internal attenuator. Instead the circuit is very often severed and the input part terminated in an external matched load at the point where the attenuator would have been. The beam is bunched by this first circuit, and after traversing a short drift space, it enters a second circuit this one being terminated at its input end in a matched load. The bunched beam excites a fresh three-wave system, of which the growing wave soon again predominates, and goes on to produce the output power. In the valve shown in Plate 9.1, which is of the clover-leaf type, the two circuits are connected to external couplings at each end. The two

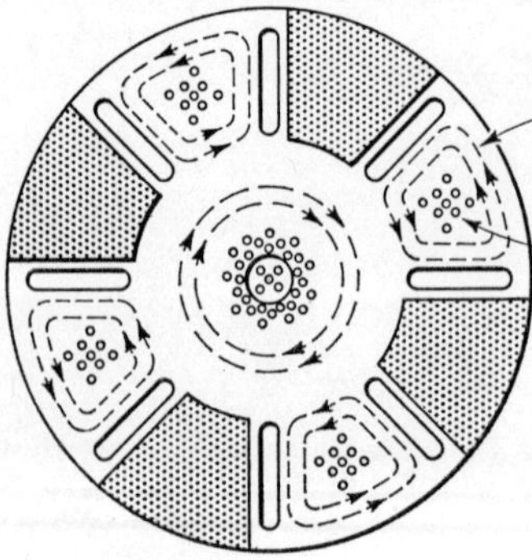

Fig. 9.19. *Principal interfering mode in clover-leaf structure*

couplings on either side of the sever can be seen on opposite sides of the valve at the centre. All these high-power tubes are, of course, cooled, usually by passing a liquid through suitable channels drilled or machined in the body.

9.9 Dispersion and Impedance

This chapter opened by mentioning dispersion and impedance, and although dispersion has been fairly thoroughly dealt with, impedance has hardly been mentioned. This is because very little can be done about it, other than ensuring that, as far as possible, electric fields are not extensive or large except where the beam is to be projected. The impedances of the circuits described are all adequately large, and in every case it is the dispersion rather than the variation of impedance that limits the valve bandwidth. The dispersion is measured by determining the resonant frequencies of a short-circuited length of the structure, and hence obtaining a series of points connecting the wavelength in the structure with the frequency, as described previously.

The coupling impedance can be measured by the application of a theorem which says that if some small perturbation is made to a resonant cavity, the fractional change in its resonant frequency f will be given by

$$\frac{\Delta f}{f} = -\frac{\delta W}{W} \tag{9.9}$$

where W is the stored energy in the cavity. In practice this is done by introducing a small piece of some dielectric material such as alumina, of known shape, into the cavity, in which the form of the field variation is known. In a slow-wave structure having a hole along the axis, for example, it is known that inside the hole the axial field variation is described by the sum of all the space-harmonic components, as indicated by the expression

$$E_z = \sum_{n=-\infty}^{n=\infty} A_n \, I_n \, (r, \theta) \, e^{j(\omega t - \beta z)} \, e^{\frac{j \, 2\pi n z}{p}} \tag{9.10}$$

Of these components, the fundamental is described by the first term of this series, and this is the only component that has a finite amplitude actually on the axis, since all the I functions (which are modified Bessel functions, whose form is well known) are zero for $r=0$ except I_0. The dielectric body is therefore made into the form of a thin rod, placed on the axis, and the variation of the resonant frequencies of the short-circuited line that this produces allows the calculation of the coupling impedance of the fundamental component. The relative amplitudes of the other space-harmonic components can be found from a Fourier analysis of the known field form at the conducting boundary, and so the coupling impedances of the space harmonics can also, with some labour, be found from measurement if desired.

9.10 Backward Wave Interaction

In the folded waveguide circuit, shown in Fig. 9.13, we have already met a circuit in which the fundamental propagation measured along the axis is of a backward-wave nature, that is, the phase change from one gap to another decreases as the frequency is increased, and it was seen that the reason for this is the geometrical phase reversal produced by the folding of the circuit. Interaction with an electron beam now takes place at a given frequency when the phase change along the beam from one gap to the next is the same as that of the wave on the circuit, *going in the opposite direction*. The situation is diagrammatically shown in Fig. 9.20, in which in the first case an electron bunch is being retarded by the field of the circuit wave. In the second case, the electron bunch has advanced one pitch, after an interval p/u, while the wave has advanced one pitch in the opposite direction, taking a time p/v, which may be

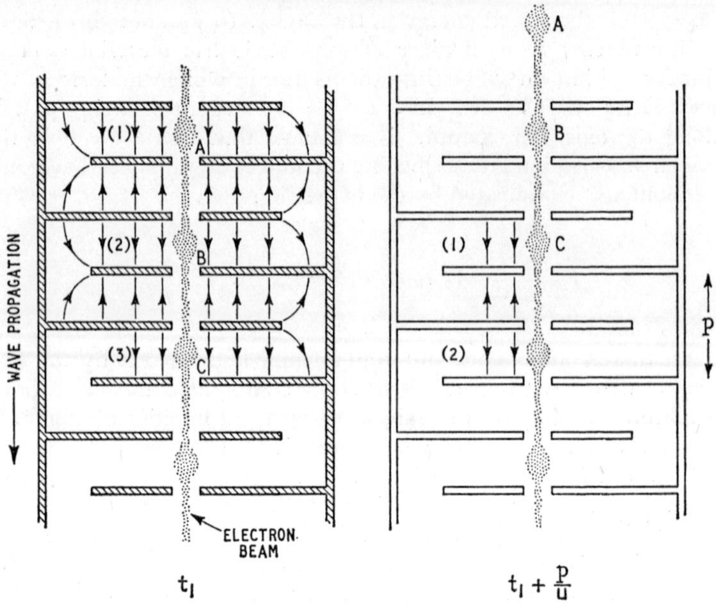

Fig. 9.20. *Backward-wave interaction in folded-waveguide*

written $p\beta/\omega$. If progressive interaction is taking place, these two must be the same, i.e.

$$\frac{\beta}{\omega} = \frac{1}{u} \tag{9.11}$$

An idealised dispersion curve for a backward-wave circuit is as shown in Fig. 9.21, in which

$$\frac{\beta p}{2\pi} = \left[1 - \frac{\omega}{\omega_0}\right] \tag{9.12}$$

If this is substituted in the above equation, we get

$$u = \frac{p}{2\pi} \frac{\omega\,\omega_0}{\omega_0 - \omega} \tag{9.13}$$

for the relation connecting the frequency at which interaction occurs with the beam velocity.

The interaction mechanism is basically similar to that occurring in a forward-wave tube (i.e. a normal travelling wave tube), but has the important difference that the wave is propagating in the opposite

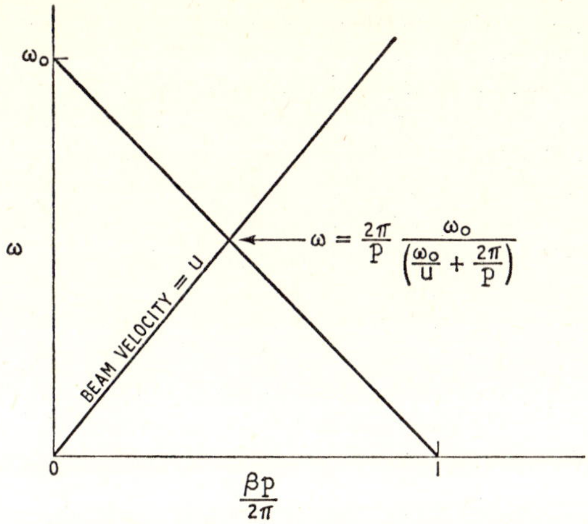

Fig. 9.21. *Idealised backward-wave characteristic*

direction to the beam, so that the wave amplitude is greatest at the end of the circuit where the beam is injected and where the beam bunching is, of course, zero, and decreases along the circuit, while the beam bunching is increasing as shown in Fig. 9.22.

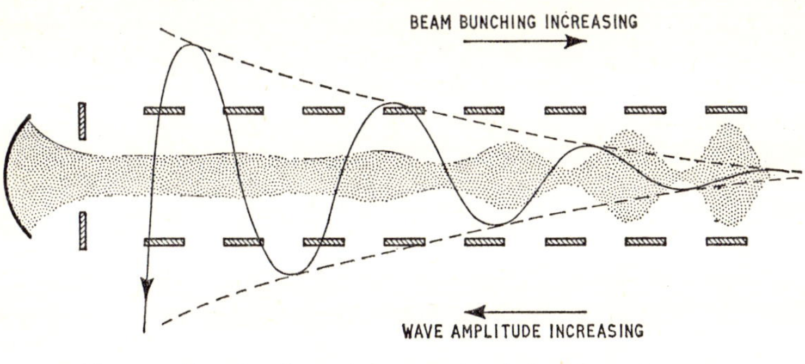

Fig. 9.22. *Beam bunching and wave growth in backward-wave interaction*

9.11 Backward-Wave Amplifiers

Returning again to the ω-β diagram, it is clear that the interaction only occurs over a narrow frequency band, of which the centre frequency is fixed by the beam velocity. If the tube is sufficiently short so that oscillation does not occur, a small signal at the correct frequency fed into the collector end of the tube will be amplified and appear at the gun end. When used in this way the device is a *backward-wave amplifier*, which is a narrow-band amplifier which can be tuned by varying the beam voltage. The maximum stable gain that can be obtained from such an amplifier using a single circuit is only about 10 dB. This is limited by the ability of the valve to oscillate, a property which also has extensive application. As the length of the circuit, or more properly the quantity CN (where C is the growing wave parameter similar to the one described in connection with forward-wave amplifiers and N is the number of electronic wavelengths along the circuit), is increased, the ratio of the input signal level to the output signal level decreases until the point is reached when CN equals about 0·3 for a lossless circuit, where the input signal required for a finite output has fallen to zero, i.e. the gain is infinite. This is the condition for oscillation to start, and the required value of CN implies that for a given current there is a critical 'start-oscillation length', or for a given circuit length a critical start-oscillation current. When used as a backward-wave amplifier the current must obviously be kept below the start-oscillation value, and the maximum gain of about 10 dB is obtained with about 0·9 of the start-oscillation current. Clearly this is not a very comfortable situation, since the start-oscillation current will vary slightly as the valve is tuned across the band, due to the variation in its electrical length and matching conditions, and if the current becomes too low relative to the start-oscillation current the gain rapidly decreases, while if too high excessive noise or even oscillation occurs. This situation is improved in practice by the use of two circuits in series along the beam, to form a cascade backward wave amplifier, shown diagrammatically in Fig. 9.23. The two

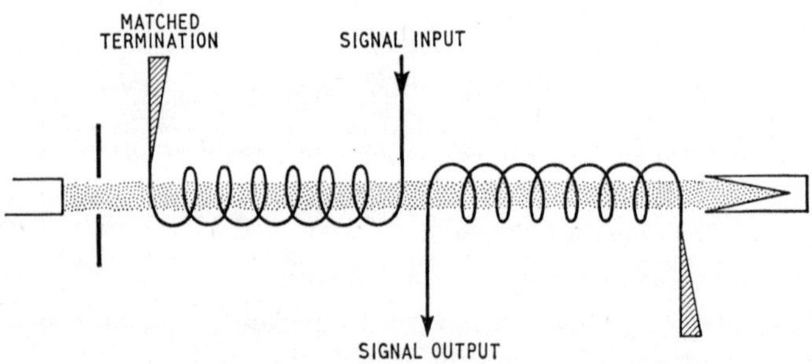

Fig. 9.23. *Arrangement of cascade backward-wave amplifier*

circuits are completely isolated from each other, like those of a severed-circuit travelling wave tube, and are coupled only by the electron beam. With this arrangement, an overall stable gain of about 20 dB can be obtained with only about 0·8 of the start-oscillation current, which has increased the margin of safety by 100 per cent. Backward-wave amplifiers have only been made for use at low power levels, since it has not been possible to make them with high efficiencies, and their principal use is as front-end amplifiers for tunable microwave receivers. For this use it is also essential for them to have the lowest possible noise figures, and, in fact, all the backward-wave amplifiers that have been made have been low-noise amplifiers, with noise figures as low as 2·3 dB. This aspect is dealt with further in Chapter 13. The demand for such valves has been fairly limited, and they have so far only been made for use in the principal microwave frequency bands, although there is no reason why they should not be used over the same range as the helix-type t.w.t.'s.

The circuit used in nearly all backward-wave amplifiers is, in fact, a helix, usually made of metal tape, which, compared with wire, improves the relative amplitude of the backward-wave space harmonic. For the same reason its diameter is considerably larger relative to the free-space wavelength than it would be in a forward-wave tube. The fact that a space-harmonic circuit rather than a fundamental backward-wave circuit is used does not, of course, affect the operation of the valve in any way.

9.12 Backward-Wave Oscillators

As the beam current of a backward-wave device is gradually increased the backward-wave gain increases until a broad noise output over the bandwidth of the tube is obtained, which is being driven by the thermal noise of the beam, and as the current is increased still further, this output becomes narrower in frequency, until it resolves into a coherent oscillation at the frequency of highest gain. The device has now become a *backward-wave oscillator*, i.e. a low-level microwave oscillator, giving an output whose frequency depends on its beam voltage in just the same way as that of a backward-wave amplifier. The valve, shown diagrammatically in Fig. 9.24, has now only one r.f. connection, at the electron-gun end of the structure, the other end being terminated in a matched section.

The backward-wave oscillator is an extremely useful device whose invention has made possible very many equipment techniques

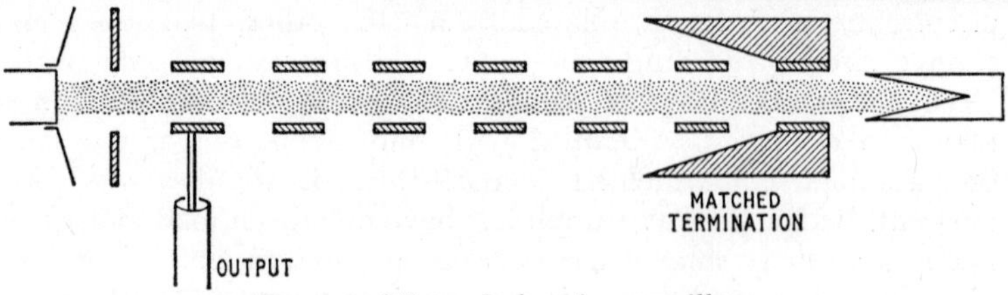

Fig. 9.24. *'O'-type backward-wave oscillator*

hitherto impossible by the use of fast electronically controlled frequency scanning. Some of these techniques are reviewed at the end of this chapter.

In a typical backward-wave oscillator, irrespective of what circuit it uses, the lowest frequency will be obtained at the lowest line voltage, the oscillation starting when the necessary value of CN is first reached. If the beam current is constant, this will also be the lowest value of beam power, and consequently the lowest level of r.f. power output occurs at this end of the tuning range. As the voltage, and hence the beam power, is increased the output power also increases. If the line voltage is also the voltage which is fixing the cathode current, the latter will also increase, and the effect will then be intensified. Most backward-wave oscillators have therefore a gross variation of output with frequency. The upper limit of frequency is set by the end of the pass-band of the circuit. In the diagrams of Fig. 9.25 are shown the voltage tuning curve, and the

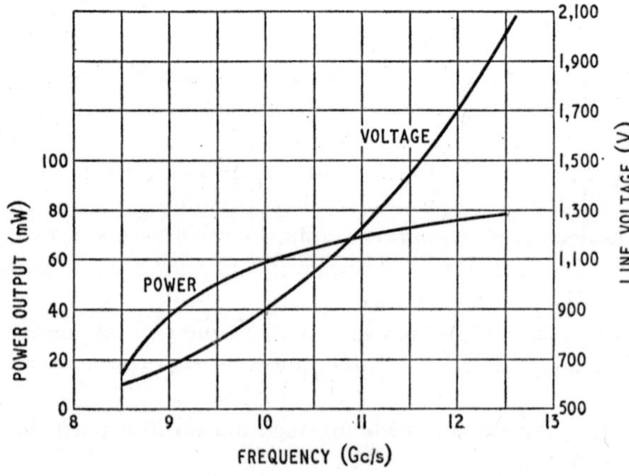

Fig. 9.25. *Backward-wave oscillator characteristics*

corresponding r.f. output power variation of a backward-wave oscillator.

In addition to the large power variation already described, which occurs at the low-frequency end of the tuning range, it is often found that there are much more rapid changes of power with frequency, and even 'holes' in the tuning characteristic at which no output is obtained. These are chiefly due to internal reflections within the tube, either at the output transition or in the matching of the line at the far end, and to achieve a smooth variation of power with frequency very great care has to be given to these two points.

The actual form and construction of backward-wave oscillators are quite varied. The rather low-power simple types for use at frequencies of a few Gc/s, used the backward-wave space harmonic of a tape helix. These had glass vacuum envelopes, with coupling similar to that used on travelling wave tubes of this construction, and were focused by external solenoids. A backward-wave oscillator is rather shorter than a travelling wave tube at the same frequency, and so it has been practical to use uniform straight permanent-magnet systems to provide the focusing field, rather than the periodic-magnet focusing systems used with travelling wave tubes which have been found to impair the behaviour of a b.w.o.

9.13 The Interdigital Line

Structures other than the helix are used, particularly at frequencies where the helix would be difficult to control in manufacture as at the very highest frequencies, or insufficiently mechanically stable as at the lowest frequencies. In these regions, the circuit usually encountered is the *interdigital line*, shown in Fig. 9.26(a). The descriptive treatment of the behaviour of this line is obviously similar to that of the folded waveguide, except that no low-frequency cut-off will now appear. Once again there is a geometric phase reversal along the axis, i.e. at a very low frequency, where the total phase change through the developed length of the circuit would be very low, there is between one gap and the next a reversal of the direction of the axial field, giving effectively a phase change from one gap to the next. As the frequency is increased to that at which the phase change along the developed circuit is π from the centre of one gap to the next, the phase change measured along the axis will gradually decrease to zero, thus giving the backward-wave characteristic shown in Fig. 9.21. From this it follows that at the highest frequencies the fingers of the line have a length of about one-quarter

(a)

(b)

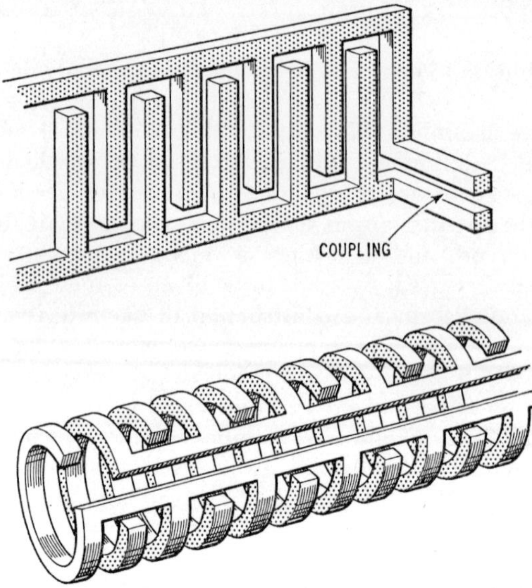

Fig. 9.26. (*a*) *Plane interdigital line* (*b*) *Cylindrical interdigital line*

of the free-space wavelength. For low-frequency valves, where this size is an embarrassment, the line may be deformed into a cylindrical shape, as shown in Fig. 9.26(b) and used with a cylindrical or annular beam, whereas in higher-frequency valves a flat strip beam is used.

9.14 The Ophitron

Another circuit which is used in connection with a very compact and convenient system of electrostatic focusing, in a device which the manufacturers call an Ophitron, consists of a ladder line mounted on a ridge-loaded waveguide, as shown in Fig. 10.18. In this valve a flat strip beam is projected along the axis close to the ladder line, and is maintained in position by a periodic electric field developed between the ladder line and the surrounding smooth electrodes, as described in greater detail in the next chapter. This circuit has a very fast forward-wave fundamental propagation, and is used in the first backward-wave space-harmonic mode. It provides an extremely compact device, having the advantage of no stray magnetic fields, and is shown in Plate 9.2.

9.15 Millimeter-Wave B.W.O.'s

At the very highest frequencies, above about 30 Gc/s (wavelength = 1 cm) the interdigital line becomes very difficult to make, and also has insufficient thermal dissipation to cope with the relatively greater beam power made necessary by increased loss per wavelength of the circuit. Such b.w.o.'s as have been made in this region have used a vane circuit, often deformed in some way, e.g. by division of the vanes by slotting, as in Fig. 9.27, to increase the

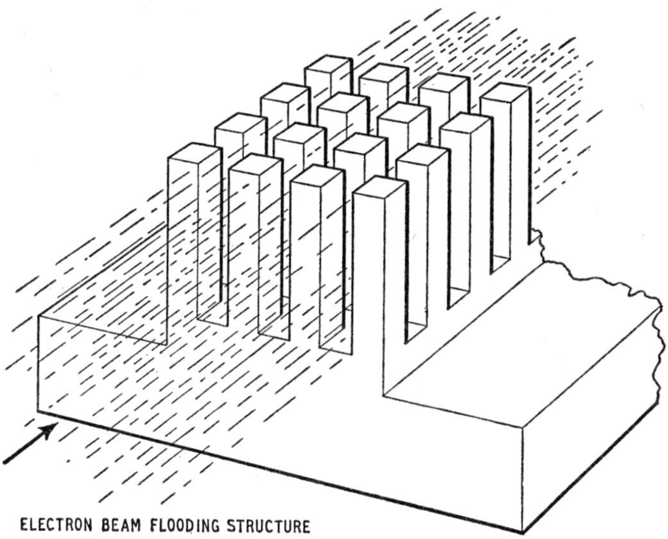

Fig. 9.27. *Slotted-vane circuit used for highest frequencies*

beam coupling. Using this approach, useful amounts of power at wavelengths down to less than 1 millimetre have been obtained, although naturally with very much less tuning range, output power and output uniformity than at lower frequencies.

9.16 Applications of B.W.O.'s

One obvious application of low-power backward-wave oscillators is in the use of frequency-swept test oscillators. At microwave frequencies circuit characteristics can be measured only by measuring the amplitude and phase of reflected or transmitted power, and if these measurements have to be made point by point, using one frequency at a time, the time required for development of any microwave circuit is onerous. By using a swept oscillator and displaying the reflected-wave characteristics on a synchronously swept oscilloscope, it is possible to obtain in a few seconds information which would otherwise have required weeks of careful tedious work.

Another use is as local oscillators in fast-tuning receivers, such as search receivers, in which it now becomes possible for the receiver to sweep very quickly through its entire band, and for example, by use of appropriate circuitry, to stop and lock instantaneously on to any selected signal. A refinement of this use is the recent development of tracked b.w.a.'s and b.w.o.'s in which low-noise frequency-selective amplification is first provided by an amplifier, while the tracked b.w.o. provides a local oscillator signal at the right frequency to give the desired intermediate frequency.

BIBLIOGRAPHY

GITTINS, J., *Power Travelling Wave Tubes*, English Universities Press (1965).

Chapter 10

FORMATION AND FOCUSING OF ELECTRON BEAMS

10.1 Parallel Flow Guns and their Limitations

In all the electron-beam microwave valves which have been described in Chapters 7, 8 and 9 we left for later consideration the questions of how the beam is obtained and formed from a cathode, and how its focusing is maintained through the valve. There is in Chapter 2 a brief discussion of the focusing action of an axial magnetic field on a thin divergent beam, in which no space-charge forces (i.e. the repulsion forces that exist between every electron and all the other electrons) were taken into account. It was shown that as long as all the electrons have the same axial velocity they will be brought to a focus every $2\pi/\omega_c$ sec, when they will form an image of the form the beam had at the point of entry to the magnetic field. A series of foci will occur along the beam, even if its speed is varying along its path. In most microwave valves the beam is a comparatively thick one, and the space-charge forces are considerable, since current densities of up to hundreds of A/cm^2 are used, and the problems involved in the design of satisfactory electron guns and focusing systems may become very difficult to solve.

Most electron-gun design starts out from the application of one of three systems in which the problem of the distribution of electric fields in the space-charge-limited electron flow has been solved. It will be appreciated that although the electric field distribution in the space between, say, two concentric cylinders is readily calculable when there is only empty space between them, this distribution will be considerably modified if the space is full of electrons, flowing under space-charge-limited conditions from one electrode to the other. Without space charge there is no divergence of the electric field at any point between the cylinders, and only one differential equation, viz.

$$\frac{d^2V}{dr^2} + \frac{1dV}{rdr} = 0 \qquad (10.1)$$

has to be solved. With space-charge flow present, the equation has to state that the divergence of electric field at any point is proportional to the space charge present at that point; but the space-charge density is itself dependent on the field distribution, and so the solution is more difficult. Three shapes for which solutions are available, however, are those of infinite plane parallel plates, infinite concentric cylinders, and concentric spheres. Taking the simplest of these first, the space-charge-limited current density (J) from one infinite plate to another opposite it, as shown in Fig. 10.1, is related

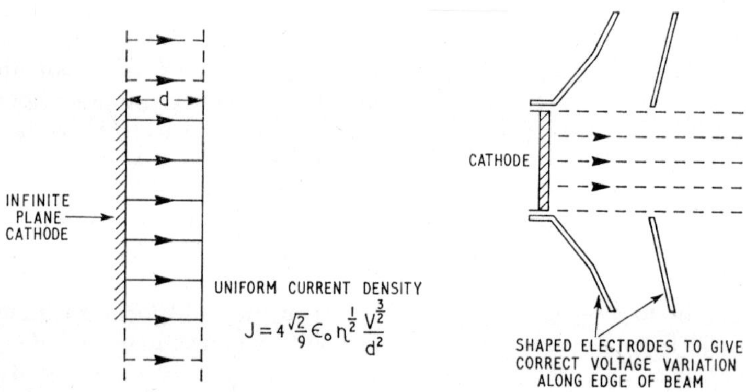

Fig. 10.1. *Infinite-plane diode and derived parallel-flow gun*

to the voltage V between them and their separation d cm by the formula

$$J = \frac{4\sqrt{2}}{9} \epsilon_0 \eta^{1/2} V^{3/2} d^{-2} \text{ A/cm}^2 \qquad (10.2)$$

and therefore the voltage variation with distance x from the cathode is given by

$$V \propto x^{4/3} \qquad (10.3)$$

This information is used in the design of an electron gun in the following way. Since the electron flow is straight and the electron paths do not cross each other, then if we consider any small circular emitting area on the cathode, this gives rise to a cylindrical beam, which is kept in place by the fields due to all the electrons around it, and which has the potential distribution given above along the direction of the beam. If now the surrounding electrons are removed (e.g. by removing the cathode coating from outside the required circular emitting area), then their effect can be simulated by suitably shaped annular metal electrodes, as shown in Fig. 10.1, which give the same potential distribution as before along the edge of the beam, and therefore the same beam flow. A particular feature of the electrode adjacent to the cathode, when it is at cathode potential, is that the angle it makes with the beam is 67·5°. This is known as the *Pierce angle* after its discoverer. The beam is then arranged to pass through a hole, which ideally should be gridded, in the opposite electrode, and can then be made use of in a valve.

This type of gun is called a parallel-flow gun, and is fairly simple to design and non-critical in its use. It is also unaffected by the presence of a uniform axial magnetic field. Its main limitation is that the emergent beam has the same diameter as the cathode, which means that the beam-current density cannot be any greater than the emission density that can be obtained from the cathode. For low-power valves, where the required beam current is small, and low-frequency valves, where the beam diameter may be large, this is acceptable, but very many valve designs require beam currents very much in excess of the maximum emission densities available from oxide-coated cathodes, which are about 0·5 A/cm^2 mean, or 5 A/cm^2 pulsed, and so some means of converging the beam must be sought.

10.2 Spherical Convergent Flow Guns

Convergent flow can be obtained by the application of the known solution of space-charge-limited flow between concentric spheres, in a manner similar to that used for parallel flow. The cathode now becomes a spherical cap, and the electrode shape surrounding this, and the shape of the anode surrounding the hole through which the beam passes, are made such that the potential profile along the edge of the beam is that which would be given by full spherical electron flow between the cathode and the anode, as shown in Fig. 10.2. The beam emerging from the anode hole is still convergent, and if left to itself would pass on to give a minimum diameter, at which, if the

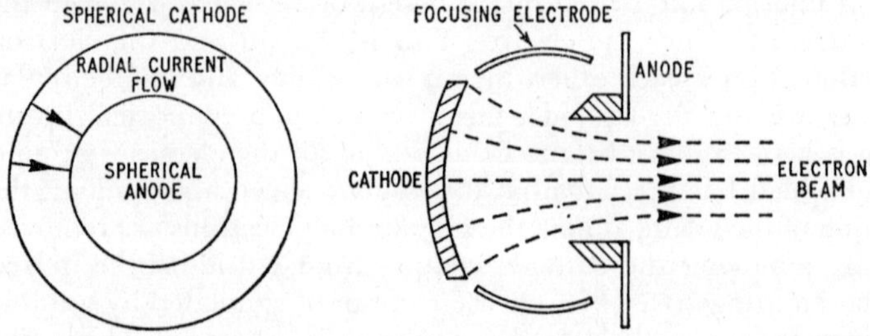

Fig. 10.2. *Spherical current flow, and the derived convergent-flow cylindrical-gun*

gun is suitably designed, all the electron flow is parallel to the axis, before diverging again under the influence of the space-charge forces. Focusing of the beam emerging from the electron gun, is however, usually provided by other means, as described later.

This type of gun is by far the most widely used in O-type valves. It provides a low cathode current density, which enables a long cathode life to be obtained, and its relatively greater size makes heater design easier. It also has the advantage that the negatively charged ions produced along the axis of the beam, which are accelerated backwards through the gun, only bombard the cathode over a relatively small area at its centre. Such ions are produced by collision of the beam electrons with molecules of residual gas in the valve, and collect along the axis of the beam. This bombardment destroys the cathode surface, and hence the emission, where it occurs.

10.3 Perveance

Under space-charge-limited conditions, for a given gun design, it is found that the quantity $I/V^{3/2}$ is constant, so that if the design is only changed by scaling all its dimensions, this quantity $I/V^{3/2}$, which is called the *perveance* (and is measured in pervs), remains the same. This is very useful in the design and investigation of guns for very small beams such as are required for millimetre wave valves, since the experimental work can be carried out on larger-scaled models.

10.3.1 Limits to Perveance of Solid Beams

The upper limit to the perveance that can be obtained from a solid-beam gun comes when the diameter of the anode hole through which the beam passes becomes comparable with the cathode-anode spacing. (Perhaps it should be said here that the electrode to which the high potential is applied, which is referred to here as an anode, since it is derived from the anode of the two electrode systems used as the basis of the design, is now also called a grid when it is used in a valve, since its function is to control the beam current, and no electrons actually flow to it except by accident.) This limit to perveance could be prevented by putting a grid across the anode hole, but this is not possible in high-power valves, which are those for which high perveances are required, because of resultant troubles with heating and evaporation of the grid, and electron emission from its wires. The maximum practical values of perveance that are used are about 2×10^{-6} pervs (or 2 micropervs), which would correspond, for example, to a beam of 2A at 10 kV, or 16A at 40 kV. The first of these might be used in a c.w. klystron or travelling wave tube having an output of 5 kW, while the second might be encountered in a pulsed valve with an output of about 200 kW. The guns would have similar electrode shapes, although more attention would obviously have to be given to avoiding voltage breakdown in the higher-voltage gun.

10.4 Strip-Beam Guns

Strip beams are used in devices where the use of r.f. circuits such as meander or interdigital lines, or ladder-loaded waveguides, makes their use desirable, and also in connection with some forms of electrostatic focusing, e.g. as in the Ophitron. Guns for these beams could be designed, at least for strips of infinite width, by the application of parallel plane flow, if no convergence is necessary, or flow between coaxial cylinders, if convergence is required.

One reason why such care is necessary in the design of electron guns for high-power valves is because of the amount of power in the beam. In a 3 mm diameter beam of say, 1 A at 10 kV, the power density is 1·39 kW *per square millimetre.* If this beam is intercepted by anything it will obviously produce intense local heating, and so the electron gun and focusing system must be designed so that the electrons do not arrive anywhere except at the collector, where their energy is allowed for. Strip beams, however, have most frequently been used in backward-wave oscillators, which are low-power devices, and consequently the beams used in them are much less potentially destructive, and so other approaches to design, in which there may be considerable interception in the gun, may be used.

One design difficulty that arises is that in practice the strip beams required are not infinitely wide, while the design procedure described is obviously based on the conception of an infinitely wide beam. This leads to difficulties at the edges and corners. For these reasons, strip-beam guns tend to be rather more complicated than cylindrical-beam guns, and designs have often been arrived at by a combination of electron optics, i.e. by calculating electron paths through the gun without considering the effects of other electrons, and experiment, and an example of a gun of this type is shown in Fig. 10.3.

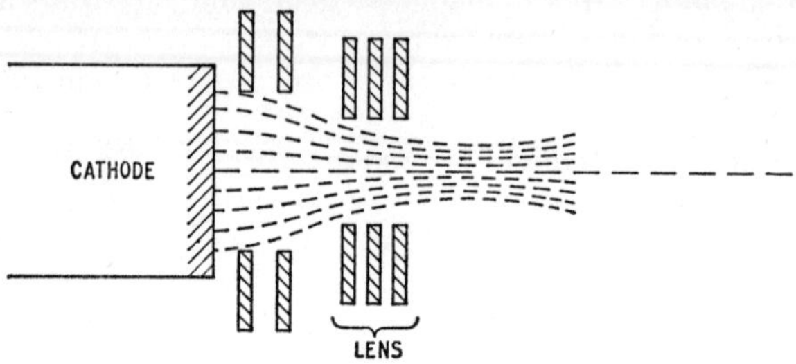

Fig. 10.3. *Strip-beam gun*

10.5 Hollow Beams

Cylindrical beams of higher perveance than can be obtained from a solid-beam gun can be produced by using a hollow-beam formation. The simplest approach to this is made by merely wrapping a strip-beam gun into a circular form. If we suppose the surface of this cathode to be divided into nearly square sectoral elements, there will be $2\pi r/t$ such elements around the cathode, where r is the mean radius, and t is the thickness. The practical limitation that now limits the maximum perveance that can be obtained occurs when the anode-cathode spacing approaches the thickness—when the perveance *per element* is greater than one microperv—so that the total perveance can be much greater. A hollow-beam electron gun of this sort, of perveance 25 μP is shown in Fig. 10.4. As in the case of solid-beam guns, it may also be desired to have some convergence in order to reduce the cathode loading, and this can be done in the same way, leading to a spherical-flow hollow-beam gun, as shown in Fig. 10.5. Alternatively, if it is more convenient, it is possible to design a gun based on a circular deformation of a convergent-strip gun as shown in Fig. 10.6.

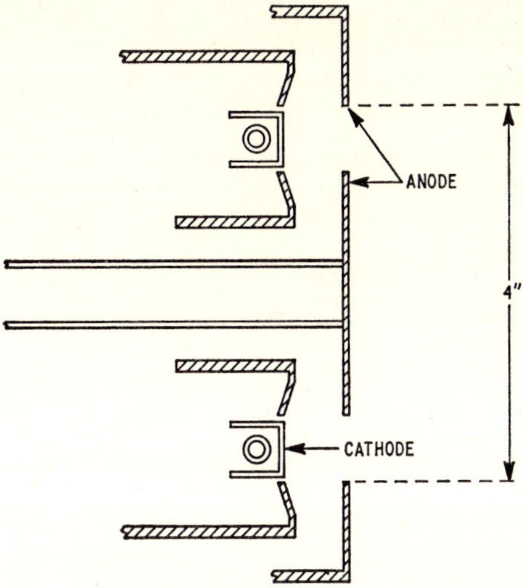

Fig. 10.4. *A* 25 μP *hollow-beam gun to give a* 4 in *diameter beam of* 25A *at* 10 kV

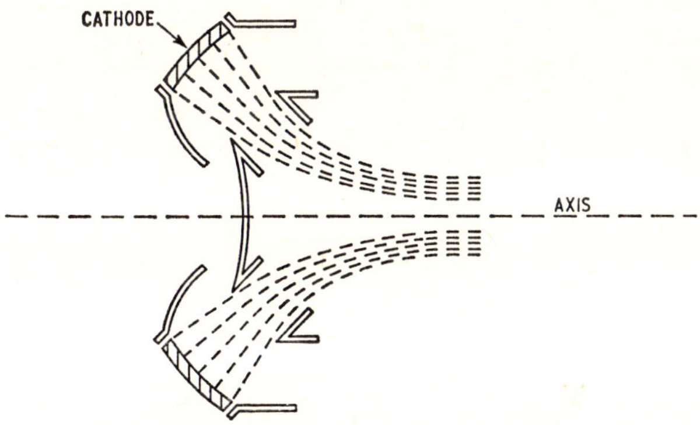

Fig. 10.5. *Spherical-flow hollow-beam gun*

10.6 Magnetron Injection Guns

There is a third approach to forming a hollow beam, which is becoming increasingly used, known as the magnetron gun. In the magnetron valve, which is described in Chapter 12, there is a cylindrical cathode, with its axis parallel to a magnetic field, surrounded by a coaxial cylindrical anode. If the magnetic field is high

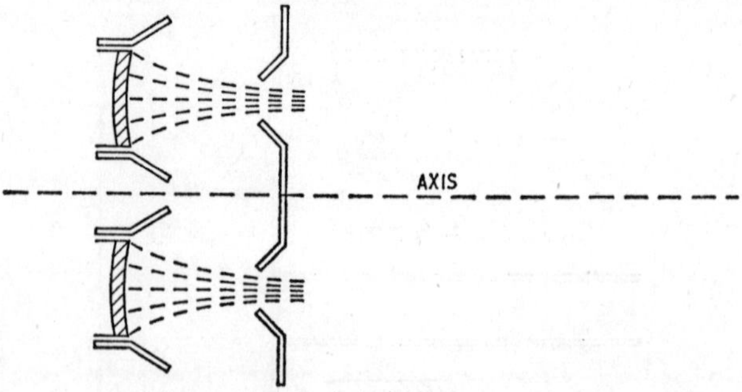

Fig. 10.6. *Hollow-beam gun derived from convergent strip-beam gun*

enough, the paths of electrons leaving the cathode are bent into a circular shape, no electrons can actually reach the anode, and a dense circulating space charge is formed in the space around the cathode. This behaviour is discussed more fully in Chapter 12. The simplest approach to a magnetron gun regards this space charge, which, like a normal cathode, has no electrons with axial velocities greater than those due to noise, as a cathode. An axial field is then provided at one end by a second anode or grid, and so a parallel beam is drawn off. The total available cathode area available is now many times the beam cross-sectional area, and so a large convergence can be obtained. Recently there has been much more detailed study of such guns, and an analysis has been made which enables the axial motion to be obtained without the complication of a second grid by suitable shaping of the cathode and anode surfaces. This gives a gun of the form shown in Fig. 10.7, in which both cathode and anode have conical surfaces. In this, we have a gun of high perveance, with a

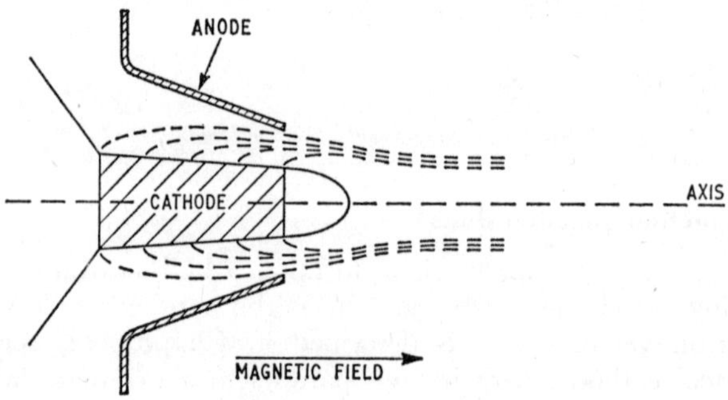

Fig. 10.7. *Magnetron injection gun for hollow beam*

very high convergence, and a relatively large spacing of the high-voltage electrodes, and at the time of writing the first valves using guns of this type are beginning to appear.

In this chapter only electron guns for O-type valves are considered. Relatively little work has been done on guns for crossed-field or M-type valves, and the description of these, and the conditions of electron flow in crossed fields are dealt with in Chapter 11.

10.7 Focusing—Practical Requirement

It has been seen that in most of the guns described the beam is electrostatically focused until it emerges from the gun. It must then be somehow held in focus during its passage through the valve. Now usually, the electrons in a beam do not behave at all in the way designers might like to think is the case, and indeed in most cases their behaviour is unknown. What this means when designing a focusing system is as follows: two things can be fixed with certainty about the beam, viz. its maximum diameter, which cannot exceed the circuit diameter, and its current, which can be measured. On the basis of these, a focusing system is designed which could focus an ideal beam in which, for example, the space-charge density is uniform and the electron flow is laminar, i.e. electron paths do not cross, and the beam is symmetrical around the axis, etc. Such beams, alas, exist only very rarely, and most frequently when the difficult and expensive work of experimental investigation has been carried out it has shown that the beam is by no means as well behaved as would be wished. Fortunately, however, the focusing systems designed for a well-behaved beam do, by and large, work with a badly behaved beam provided that, for example, the magnetic fields used are not fixed at the minimum theoretical value, but generously applied. In fact, the only things asked of the focusing system are that it shall effectively get the beam through the valve without too much current being intercepted on the circuit, and that the beam shall be effective in doing its job and not be so disordered by the focusing fields that it cannot provide continuous r.f. interaction.

10.8 Theory of Ideal Focusing

Most of the theoretical treatment of the focusing of beams on the other hand is, in order that it shall be simple enough to enable useful results to be reached, carried out on the assumption that the beam is more or less an ideal one. It should not be concluded from this that ideal well formed beams are never obtained. They have often been produced and demonstrated experimentally in beam-analysing experiments, and have been focused very much as theory would predict.

From the foregoing, it will be apparent that a full analysis of beam behaviour cannot be given here. We can, however, look at some of the ideal behaviour referred to. Let us suppose that we have a homogeneous cylindrical beam, symmetrical about its axis, flowing so that its electron paths are not crossing, in a uniform magnetic field of flux density B parallel to the beam axis. Quite a lot of assumptions. First, we should note that if $\bar{\rho}$ is the mean charge density inside a radius r, then the radial electrical field E_r at this radius is given by Gauss's theorem, which says that the total integrated field perpendicular to the surface around a volume enclosing a charge is equal to the charge enclosed divided by the permittivity. In this case, because of the symmetry, we can consider a unit length of beam, as shown in Fig. 10.8, for which we may then write

$$2\pi r E_r = \frac{\pi r^2 \bar{\rho}}{\epsilon_0} \tag{10.4}$$

Hence the radial electric field acting on electrons at a radius r is given by

$$E_r = \frac{r\bar{\rho}}{2\epsilon_0} \tag{10.5}$$

The forces then acting on an electron at this radius are:

(1) that due to the space-charge field acting outwards, eE_r,
(2) that due to the magnetic field and the azimuthal motion (of angular velocity ω), acting inwards, $Ber\omega$, and
(3) that due to the magnetic interaction between the field and the radial motion, acting in a tangential direction, $Be\dfrac{dr}{dt}$.

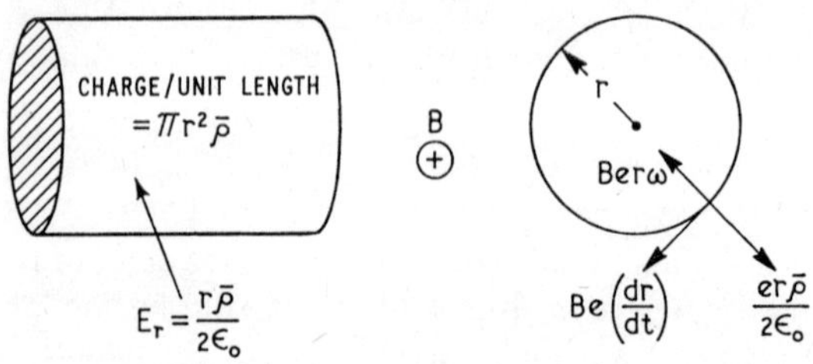

Fig. 10.8. *Radial field due to space-charge, and forces on an electron*

The radial equation of motion of the electron is therefore

$$\frac{er\bar{\rho}}{2\epsilon_0} - Ber\omega = m\left(\frac{d^2r}{dt^2} - r\omega^2\right)$$

i.e.

$$\frac{\eta\bar{\rho}r}{2\epsilon_0} - \eta Br\omega = \frac{d^2r}{dt^2} - r\omega^2 \tag{10.6}$$

By examination, it can be seen that if

$$\frac{d^2r}{dt^2} = 0 \text{ and } \frac{dr}{dt} = 0$$

in which case the beam is in equilibrium with a smooth cylindrical boundary, then Equation 10.6 becomes

$$\frac{\eta\bar{\rho}r}{2\epsilon_0} - \eta Br\omega + r\omega^2 = 0 \tag{10.7}$$

If we divide throughout by r, and write ω_p^2 for $\frac{\eta\bar{\rho}}{\epsilon_0}$ and ω_c for ηB (see Chapter 6), this becomes

$$\frac{\omega_p^2}{2} - \omega_c\omega + \omega^2 = 0 \tag{10.8}$$

This will give a constant value of ω throughout the beam if ω_p is constant, i.e. if $\bar{\rho}$, the mean space-charge density is constant, so that the beam is then rotating as if it were a solid rod.

The angular velocity ω is also given by a well-known theorem, Busch's theorem, which states that if a thin cylindrical hollow beam of electrons with radius r leaves a cathode through which the magnetic flux is φ_0, its angular velocity at any later time, at a point where the flux through the beam is φ is given by

$$\omega = \frac{\eta}{2\pi r^2}(\varphi - \varphi_0) \tag{10.9}$$

If the field is uniform and parallel to the beam, then $\varphi = \pi r^2 B$ so

$$\omega = \frac{\eta B}{2}\left(1 - \frac{\varphi_0}{\varphi}\right) \tag{10.10}$$

Let us also suppose that ρ, the space-charge density, is constant throughout the beam, so that $\bar{\rho}$ the mean density inside a radius r is also constant, and that φ_0 the magnetic flux through the cathode is zero, so that

$$\omega = \frac{\eta B}{2} = \frac{\omega_c}{2} \tag{10.11}$$

Equation 10.8 then becomes

$$\frac{\omega_p{}^2}{2} - \frac{\omega_c{}^2}{4} = 0 \tag{10.12}$$

i.e.

$$\omega_p{}^2 = \frac{\omega_c{}^2}{2} \tag{10.13}$$

What has been done is to show that if this condition is fulfilled, then a beam of constant space-charge density can be held focused in a perfectly parallel condition. Such a beam might be launched from a suitably designed gun, which is completely shielded from the magnetic field in order to fulfil the $\varphi_o = 0$ condition, into a region of uniform field as shown in Fig. 10.9. This particular flow, known as

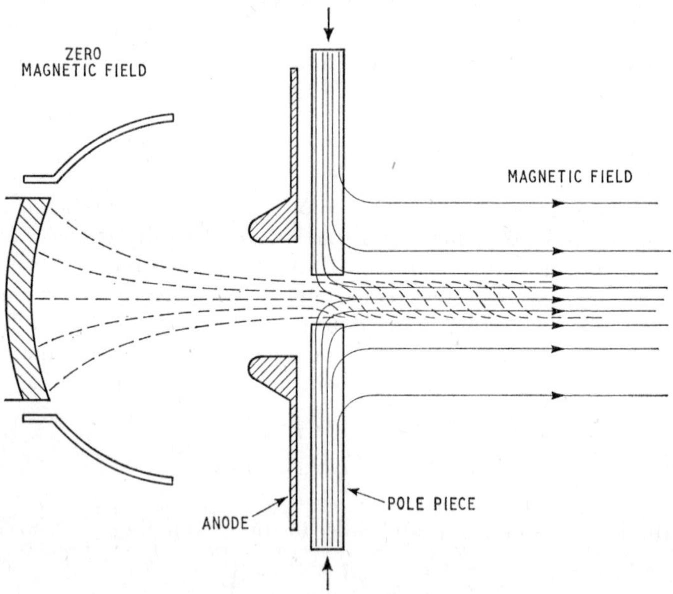

Fig. 10.9. *Injection into Brillouin field*

Brillouin flow, only occurs for one particular value of magnetic field, called the Brillouin field, defined by Equation 10.13, which may be rewritten.

$$B^2 = \frac{2\rho}{\eta \epsilon_o} \tag{10.14}$$

This is the only field, other than an infinite field, which will produce completely parallel flow, and, as stated previously, by careful gun design and attention to entry conditions, it has been demonstrated experimentally. The Brillouin beam has constant charge density and angular velocity throughout and is thus rotating as a solid rod. It is of interest to know the current density distribution, and to find this we need the axial velocity distribution. There will be, as in any other sort of beam, a reduction of potential in the centre of the beam relative to the outside, due to the space-charge field. It was noted in Equation 10.5 that the radial electric field was

$$E_r = -\frac{dV}{dr} = \frac{\rho r}{2\epsilon_0}$$

hence the space-charge potential reduction, obtained from integrating this gives

$$V = V_o + \frac{\rho}{4\epsilon_o} r^2 \qquad (10.15)$$

for the potential variation across the beam. The velocity of an electron at any point is made up of its components v_z, parallel to the axis, and $v_\phi = r\omega$ due to the beam rotation. Writing the energy equation for this case, therefore, gives

$$[\tfrac{1}{2}mv_z^2] + [\tfrac{1}{2}m\omega^2]\, r^2 = eV_o + \left[\frac{\rho e}{4\epsilon_o}\right] r^2 \qquad (10.16)$$

Since both sides have the same form, we can equate terms, which gives the result that the axial velocity $v_z = [2\eta V_o]^{\frac{1}{2}}$ and is constant throughout the beam, while the variation of potential due to the space-charge reduction is exactly compensated for by the variation of energy due to the rotation. It must be noted that V_o here is not the voltage by which the beam has been accelerated, but only that part of it which has produced the axial motion. The total voltage to produce this depends on the space charge density and the beam radius, as given by Equation 10.15.

It will be appreciated that this is an exceedingly neat solution, in which everything seems to be ideally simple. The reasons why it is not used in practice, apart from the difficulty of setting it up, are firstly because of r.f. bunching, which may increase the local charge density to many times that of the unbunched beam, and secondly because of the higher noise output which is obtained if the cathode is magnetically shielded.

The Brillouin field is useful, however, as a reference field, and its calculation, which is relatively simple, gives a good guide to the magnetic field, usually about 1·5 to 2 times the Brillouin value, which will actually be required to focus a given electron beam.

10.9 Magnetic Fields

It has already been said (Chapter 2) that the magnetic field does no work in affecting the electron motion, and hence in principle no continuous supply of energy should be needed to maintain the field. A solenoid, however, which is the device by which a uniform magnetic field is usually obtained, needs a considerable amount of power in order to function, merely in order to overcome its d.c. resistance, and all this appears as heat, which must be allowed for. The power required increases as the square of the field, and also increases as the size of the solenoid decreases, so that if the solenoid is made small, and therefore cheap, it becomes much more expensive to run and cool. It is therefore an undesirable device.

10.9.1 Use of Permanent Magnets

Why, then, does one not use a permanent magnet, as is done in magnetrons and crossed-field backward-wave oscillators? The answer lies in the nature of the field required. In valves such as klystrons, O-type backward-wave oscillators and travelling wave tubes, which are long-beam devices, the magnetic field requirement is for a long thin region of field, with fair uniformity of field strength along its length, but with extremely high uniformity of direction. The actual magnitude of field required is not usually very great, typically in the region 200 to 600 oersteds, which is no greater than the coercive force of the most commonly met magnetic materials. The sort of magnet that would be used therefore would be essentially a straight cylinder with the valve inside it, lying along its axis. The magnetic field configuration given by such a magnet is shown in Fig. 10.10, where it can be seen that there is along the axis a region of field that apparently has the required characteristics. In order to obtain a uniform field, it becomes necessary to deform the cylinder into an ellipsoidal shape, as is shown in Fig. 10.11.

If it is attempted to focus a valve such as a travelling wave tube in this sort of magnet, however, it is often unsuccessful. This is because the magnetic material, which is usually a very hard cast alloy, is insufficiently uniform in composition to give a perfectly straight magnetic axis, although it may be very well adjusted and magnetised so that its axial field component is exceedingly uniform.

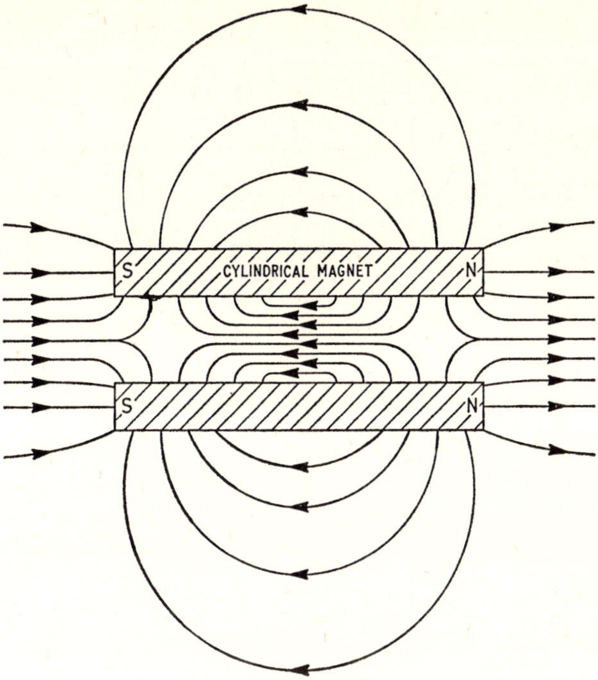

Fig. 10.10. *Field of cylindrical magnet*

This is not easy to detect by measurement with a fluxmeter; the easiest way to discover defects is probably by attempting to focus an electron beam. Such defects can be remedied by the use of a number of thin ferromagnetic discs, appropriately called field straighteners, placed inside the magnet, as shown in Fig. 10.11. These discs are virtually transparent to axial fields, but any radial field components will concentrate into them, leaving only the axial field components in the spaces between them.

A second disadvantage to the use of straight-field magnets is the amount of space that they require. Referring again to the diagram of Fig. 10.10 it is seen that there is a considerable region outside the magnet over which its field extends. This space must be kept free of all magnetic materials since their presence would distort the field not only externally but internally as well. This is very inconvenient, because where microwave valves are used space is often at a premium. Nor is it easy to remedy this by magnetic screening, as is done

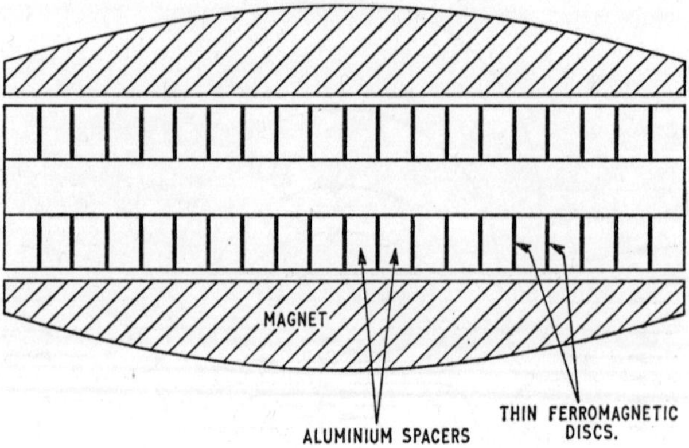

Fig. 10.11. *Field straighteners in a straight-field ellipsoidal magnet*

with chokes and transformers, because the field produced by the magnet in the region where it is desired depends on the total reluctance of the space between its poles. If a substantial screen is placed close to the outside of the magnet, this acts as a magnetic short circuit, and the field everywhere else is reduced. This must in turn be compensated for by increasing the magneto-motive force of the magnet by increasing its size, when it requires more screening, etc. and it may not be possible to find an acceptable compromise. The situation is roughly analagous to having to flood a room having drains at its edges, in order to obtain a given depth in the required area in the centre. If it is inconvenient to have the water everywhere else, drains can be provided around the centre, which then needs an even bigger supply to maintain the level.

10.9.2 Periodic Permanent Magnet (P.P.M.) Focusing

A very great improvement on this situation is provided by the idea of *periodic magnetic focusing*. This may be simply thought of in the following way, If an electron beam is shot through a short magnetic circuit, as indicated in Fig. 10.12, the circuit acts as a converging lens, and in the absence of space charge the beam would be brought to a focus farther along the axis. This behaviour, which is often used to focus cathode ray tube beams, will be well known. When the space charge is considerable, as in most of the beams of interest in microwave valves, then instead of producing a focus, the incident beam, which will already be diverging because of the space charge,

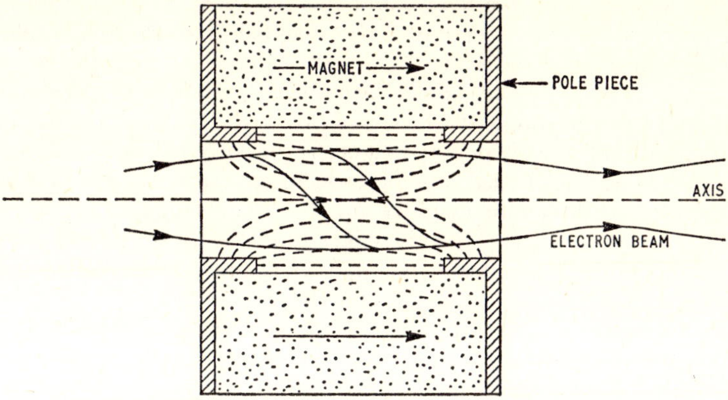

Fig. 10.12

will be made convergent. After passing through the lens the space-charge forces will again be predominant, and the beam will again start to diverge. It can then pass into a second, similar lens, with a similar effect, and then a third, and so on, giving finally a series of lenses providing a rippled beam along their axis, as shown in Fig. 10.13. The converging action of the lens is independent of the direction of the magnetic field. If the beam is not rotating before its entry it commences to rotate as it enters the field, interaction between the axial field and the rotating beam gives it the convergence, and the rotation is removed again as it passes through the radial field region on leaving the lens. The direction of the field affects only the direction of the rotation. It is thus possible to alternate the direction of

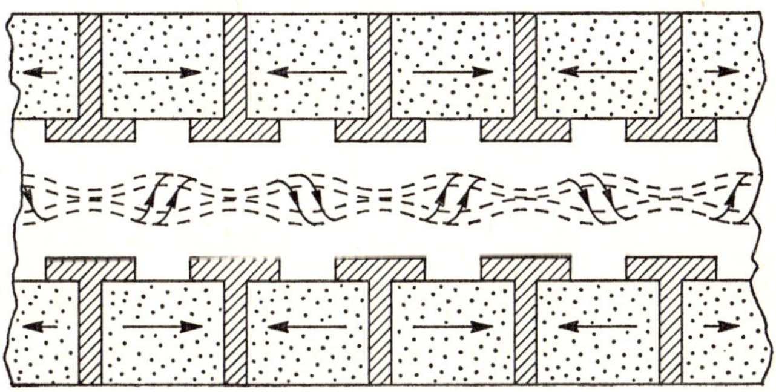

Fig. 10.13

the field in successive lenses, and this has an extremely important consequences on the design of the magnetic circuit because of its effect on the leakage fields.

Much of the material necessary in a straight-field permanent magnet goes to supply the leakage fields in the space around it. The leakage field of a single bar magnet is proportional to its magnetic moment. If it is divided into n pieces, and alternate lengths are then reversed in direction, to make up the periodic magnet shown in Fig. 10.13, the leakage field is reduced by a factor of approximately n^2, and so the amount of magnetic material needed can be reduced by a similar factor.

A detailed analysis of the electron motion shows that it should be possible to obtain no scalloping of the beam by using a maximum field equal to the Brillouin value, and a suitable period. In practice, as in the case of straight-field focusing, it is necessary to use a higher field, e.g. of about twice the Brillouin value. A system of period L cm will then focus electron beams at voltages above some critical voltage V_c given by

$$LB > 15{\cdot}3\ V_c$$

where B is the magnetic flux density in gauss.

This type of focusing, together with the use of coupled helix couplers, makes possible compact and simple solutions to the joint problems of focusing and coupling in travelling wave tubes as shown in Fig. 8.15.

At higher frequencies, where waveguide couplings are used, the periodic magnetic field must be maintained across the gap caused in the magnetic circuit by the waveguide. One effective way of doing this, by the use of a magnetic circuit going round the waveguide, is shown in Fig. 10.14.

10.9.3 Choice of Magnetic Materials

Usually, the magnetic field strengths required are no greater than the coercive force (about 1,000 oersteds) of common magnetic materials such as sintered barium ferrite and the arrangement used is then as shown in Fig. 10.13. If the fields are low enough (e.g. below about 500 oersteds) it is possible to use nickel-iron-based magnets, which have larger stored energies, and hence need less magnetic material for a given magnetic field. For fields above about 1,000 oersteds it is necessary to use more specialised and expensive materials, such as domain-oriented barium ferrite, or even platinum-

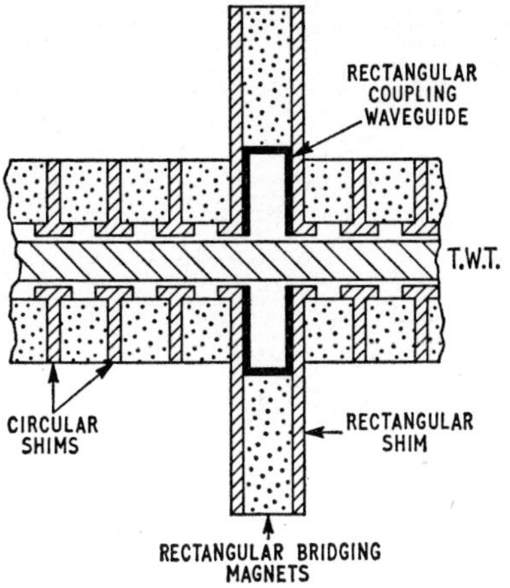

Fig. 10.14. *Maintenance of periodic field across a waveguide-coupling gap*

cobalt alloy. The values of $(B\ H)_{max}$ which govern the volume required, and H at the $(B\ H)_{max}$ point are listed in Table 10.1.

Alternatively it may be desirable to depart from this circuit arrangement in favour of a system which permits the length of the magnet in its magnetised direction to be greater than the length of the field it produces. One example of such a circuit is shown in Fig. 10.15. Such an arrangement, although much larger than the axially

Table 10.1 MAGNETIC PROPERTIES OF SOME TYPICAL MAGNET MATERIALS

Material	*Energy Product* $(BH)_{max}$ *kilogauss oersteds*	*H at* $(BH)_{max}$ *oersteds*	H_c *oersteds*
Isotropic Barium Ferrite e.g. Magnadur I, Feroba I, Indox I	$1{\cdot}0 \times 10^6$	1,000	1,750
Oriented Barium Ferrite e.g. Magnadur II, Feroba II	$2{\cdot}5 \times 10^6$	1,200	2,500
Sintered Nickel-Cobalt-Aluminium e.g. Alcomax II	$13{\cdot}0 \times 10^6$	500	580
Hycomax II	$3{\cdot}8 \times 10^6$	750	1,200
Ticonal G	$13{\cdot}0 \times 10^6$	500	580
Platinum-Cobalt	$9{\cdot}0 \times 10^6$	2,750	4,800

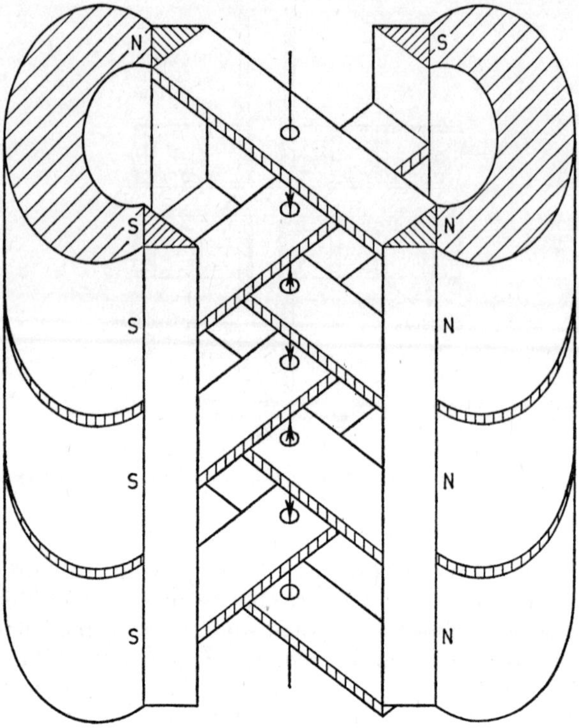

Fig. 10.15. *Periodic focusing system for high fields*

magnetised system, is still greatly preferable to either a solenoid or a straight field system.

10.10 Electrostatic Focusing

So far, all the focusing systems described have been magnetic. Various electrostatic systems have also been described in the literature, but few of these have proved reliable enough in practice. Two exceptions to this are those of the Estiatron (R.C.A.) and Ophitron (G.E.C.). Both of these use periodic electrostatic lenses to maintain the beam, the general principle being similar to that of periodic magnetic focusing. The important difference that arises in periodic electrostatically focused valves is that the periodic field is formed by the r.f. circuit, which now has a dual role.

In the Estiatron (Greek estia—meaning focus) the usual helix of a travelling wave tube is replaced by two similar helices, interwound in the same direction. These are held at different d.c. potentials, of which one is considerably higher and the other lower than the effective beam voltage (Fig. 10.16). The stationary d.c. fields of

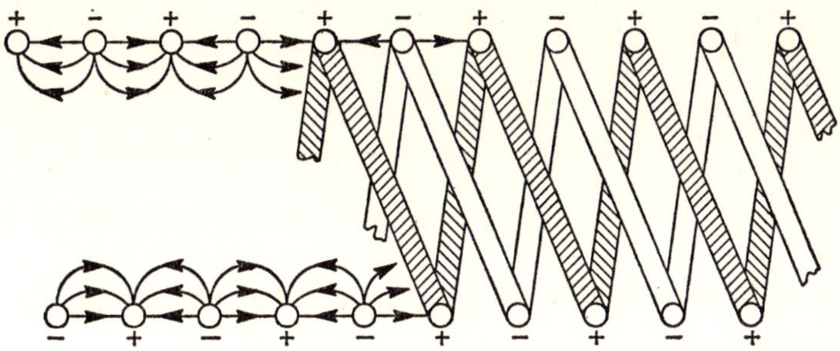

Fig. 10.16. *Electrostatic fields of bifilar helix (arrows show directions of forces on electrons)*

these helices, which have the form shown in the figure are, like those of the periodic magnetic system, alternately radial and axial. The axial components will alternately slow down and speed up the outer electrons of the beam, while the radial components will deflect them alternately outwards and inwards. If we consider the path of an electron near the outside of the beam, as shown in Fig. 10.17, it can

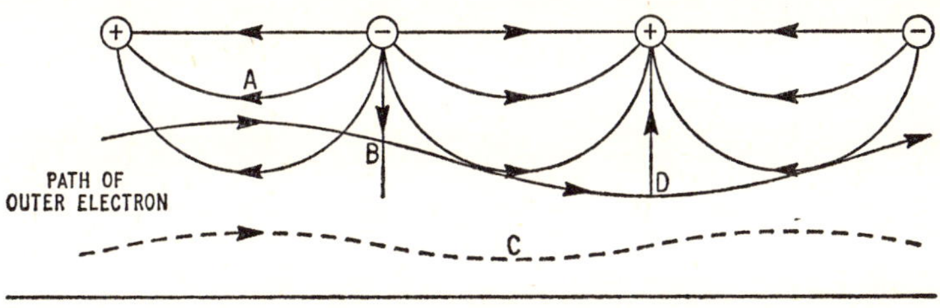

Fig. 10.17. *Focusing action of bifilar helix (arrows show directions of forces on electrons)*

be seen that in the region *A* the electron is being slowed down, while at *B*, while passing through the radial field which is considerably greater than the outward space-charge field, it is deflected towards the axis. At *C* it is speeded up, and therefore passes more quickly through *D*, where it is deflected away from the axis. In addition to spending longer at *B* under the action of the inward force, the electron is at this time closer to the circuit, so that this force is stronger than the outward force at *D*. It will be clear also that the whole action will be strongest for outer electrons, where the space-charge fields are also strongest, and zero at the centre where in principle no focusing is necessary. The overall effect therefore is to prevent spreading of the beam.

The resulting valve looks just like a comparable solenoid-focused t.w.t., but, of course, is complete in itself. The two helices are excited in phase by the r.f. input coupler, when their action is very similar to that of a single helix.

A difficulty that arises in the design of this valve is that of avoiding d.c. breakdown between the two helices, which are very close together and must be somehow supported along their length. There is thus a very long narrow breakdown path in the valve. At the cathode end this path may become coated by the barium which is continuously evaporating from the cathode. This deposition is a hazard present in all types of valve, but in an electrostatic bifilar helix system its effects are much more serious.

10.10.1 The Ophitron

The Ophitron (Greek ophis—a serpent) shown in Plate 9.2 is an electrostatically focused backward-wave oscillator, in which a ribbon beam is used with a loaded ladder-line circuit. The circuit, shown in Fig. 10.18(a) is essentially that of a ridged waveguide, in which the plane face is replaced by a ladder. Its r.f. characteristics are described in Chapter 9.

The beam focusing is effected by the d.c. fields obtained by placing a smooth electrode above the ladder line, and holding this electrode, the line itself, and the top of the ridge at different d.c. potentials. The field variation that this gives over a cross-section along the axis of the valve is shown in Fig. 10.18(b). Below the rungs of the ladder, the steady field is always in the same direction, shown by the upward arrows, but in the space between the ladder and the focusing plate the direction of the field alternates as shown, provided the spacing between the rungs is at least as large as the spacing between the ladder and the focusing plate. These fields are transverse to the electron motion and so the electrons are deflected to and fro, as shown by the dotted line, into a serpentine path which provides the inspiration for the name.

As in all b.w.o.'s the r.f. line is carefully matched by a resistive termination at the collector end, and the output is brought through a coaxial connection at the gun end.

The application of this type of focusing is particularly appropriate

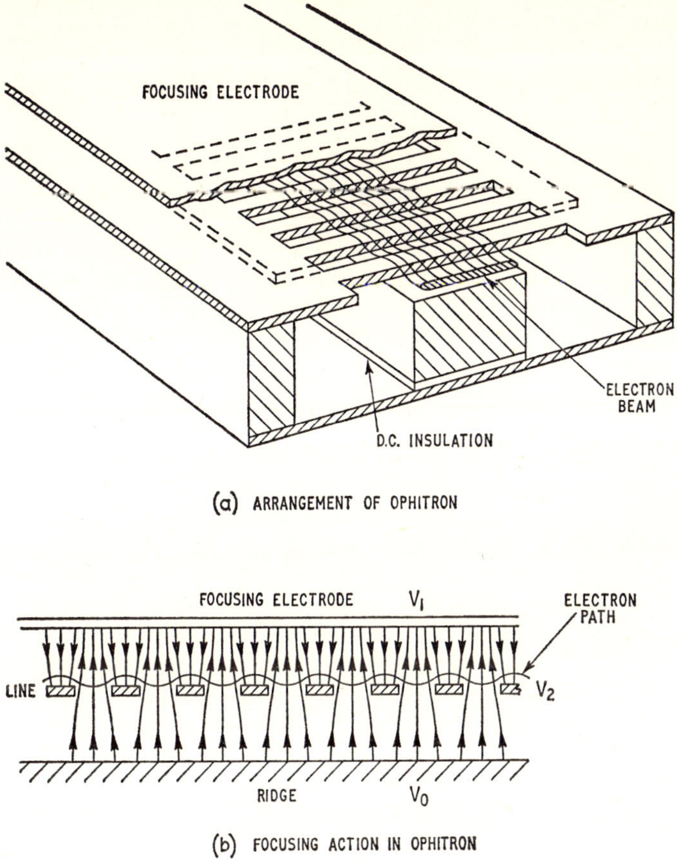

Fig. 10.18 (*a*) *Arrangement of ophitron.* (*b*) *Focusing action in ophitron* (*arrows show directions of forces on electrons*)

in the case of a b.w.o., since, owing to the fact that the interaction is with a space-harmonic having an azimuthal variation, it is not possible to use periodic magnetic focusing, which violently affects azimuthal velocities. The transverse statis fields also serve to give very rapid ejection of ions from the beam, which is important in the avoidance of unwanted noise modulation.

Some of the characteristics of the Ophitron are shown in Fig. 9.25.

BIBLIOGRAPHY

PIERCE, J. R., *Theory and Design of Electron Beams,* Van Nostrand (1954).

Chapter 11

LINEAR CROSSED-FIELD VALVES (M-TYPE VALVES)

11.1 Types of Crossed-Field Valves

Crossed-field valves differ from beam-type valves in several ways. The electrons are drawn from a cathode, but are in most cases largely intercepted by the circuit. They give up energy to the circuit, but without slowing down, and they are formed into r.f. bunches, but without necessarily increasing the space charge. These are the two aspects of behaviour which cause the limitation of power and efficiency in beam-type valves. Hence it is mainly for efficient high-power operation that the crossed-field arrangement is used.

The essential features of all crossed-field valves are indicated in Fig. 11.1. There is always a r.f. circuit, whose elements lie in a plane or cylindrical surface, opposite a smooth conducting surface, which may be the cathode, but is often a non-emitting electrode called the sole. A large steady electric field is applied between the two electrodes. There is also a magnetic field at right-angles to both the axis of the circuit and to the electric field. Unlike the situation in travelling wave tubes where the magnetic field does not enter into the r.f. interaction, in crossed-field valves this magnetic field plays an essential part in the operation. The electron beam is usually of rectangular cross-section, with the E field across its narrow dimension and the B field parallel to its width.

The motion of an electron in the first system of Fig. 11.1 was described in Chapter 2, where it was seen that the mean axial velocity was always E/B, and that various cycloidal types of tra-

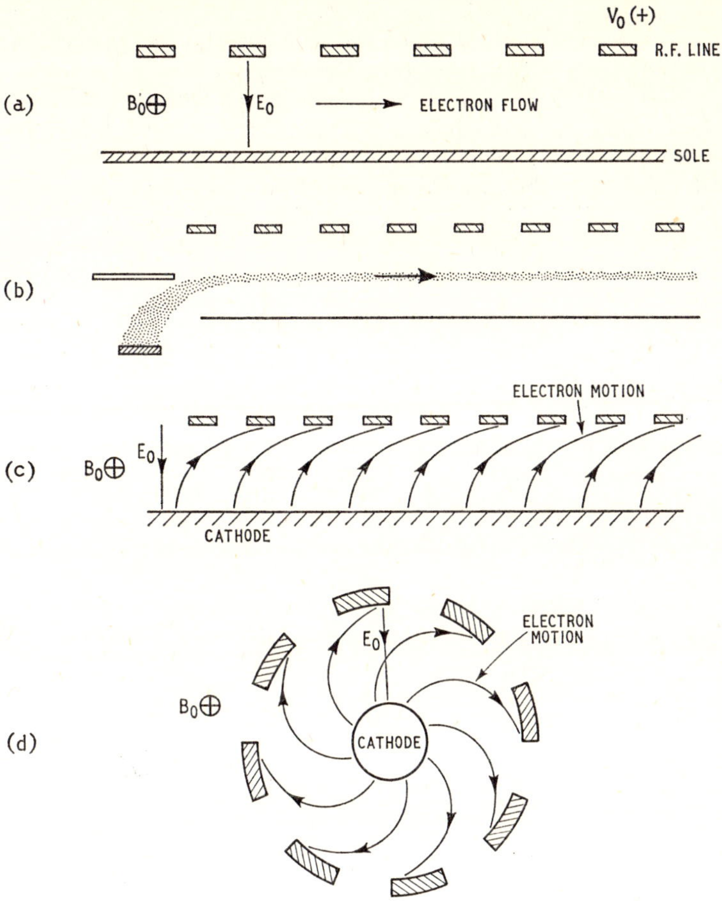

Fig. 11.1. *Crossed field arrangements: (a) general crossed-field arrangement (b) linear injected beam (c) linear continuous-cathode (d) cylindrical re-entrant beam*

jectory were possible, in which the cycloidal frequency was ω_c, the appropriate cyclotron frequency. The actual trajectory of any electron would depend on its initial velocity.

In some valves, such as the crossed-field backward-wave oscillators (often called M-Carcinotrons, although strictly this title only refers to valves made by the French C.S.F. Company), the beam is injected into the interaction space from an electron gun outside it.

These are called linear injected-beam valves. In actual practice, at the lower frequencies, the linear circuit is bent into a circular form in order to gain compactness, but this does not affect the conception. In other valves, the cathode may extend right through the interaction region, so that fresh current is supplied as fast as it is removed by the line, while in cylindrical systems, such as the Magnetron and Amplitron, both the cathode and the electron beam are continuous right round the inside surface of the cylindrical interaction space as in Fig. 11.1 (d).

11.2 Electron Beams in Crossed Fields

Since the mean velocity of all electrons must be E/B it is possible to get laminar flow, i.e. all the electrons flowing smoothly, only when they are all injected into the system at this velocity. The steady electric field E_o is produced by a voltage V_o applied between the sole and the line and hence $E_o = V_o/d$. The voltage V_y at any plane at a distance y from the sole is then

$$V_y = \frac{V_o}{d} \cdot y \tag{11.1}$$

To arrive with the right velocity E/B, each electron would have to have been accelerated through a voltage difference V given by

$$\tfrac{1}{2}m \left(\frac{E}{B}\right)^2 = eV \tag{11.2}$$

The cathode voltage V_c of the layer of electrons entering the system in the plane at y would therefore have to be given by

$$\frac{V_o y}{d} - V_c = \frac{1}{2\eta} \cdot \left(\frac{V_o}{Bd}\right)^2 \tag{11.3}$$

Hence such a beam would have to come from a large number of cathodes, each at the right voltage to give the desired velocity. This is not practical.

11.2.1 Laminar Flow in Crossed Fields

It is, however, possible to have laminar flow with a beam from a single cathode if the space-charge density is high enough. The electric field due to the space charge then modifies the potential variation across the beam so that it is just right to give the electron velocity met at any point. Such a beam therefore has a considerable

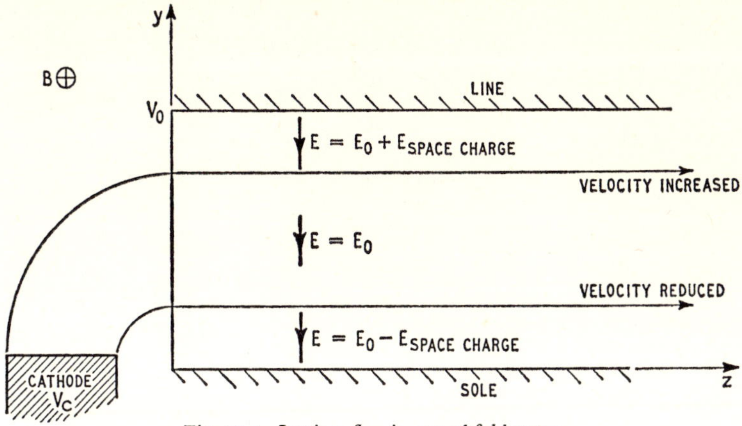

Fig. 11.2. *Laminar flow in crossed-field system*

velocity variation across it, with the electrons nearest the line going much faster than those nearest the sole (see Fig. 11.2).

As always, in the space between the line and sole, the divergenec of the electric field is given by Poisson's Equation

$$\frac{\partial E_x}{\partial x} + \frac{\partial E_y}{\partial y} + \frac{\partial E_z}{\partial z} = \frac{\rho}{\epsilon_0} \tag{11.4}$$

In the present case, taking axes as shown, we assume that there is no variation of field in the x or z directions, and that the field in the y direction is that due to the voltage V_0 applied between the electrodes plus the field due to the space charge. If there is laminar flow, and all electrons have velocities only in the z direction, electron accleration and velocity in the y direction must be zero. We also know that the electric field has components only in the y direction, so that $\frac{\partial E_x}{\partial x}$ and $\frac{\partial E_z}{\partial z}$ are zero. The above equation thus reduces to

$$\frac{\partial E_y}{\partial y} = \frac{\rho}{\epsilon_0} \tag{11.5}$$

We also know that in such flow the electron velocity at every layer is given by

$$v = E_y/B \tag{11.6}$$

hence by differentiating,

$$\frac{\partial v}{\partial y} = \frac{1}{B}\frac{\partial E_y}{\partial y} \tag{11.7}$$

Now both E_y and v are varying in ways which depend on each other, and which we so far do not know, but by combining Equations 11.5 and 11.7 we get

$$\frac{\partial v}{\partial y} = \frac{\rho}{\epsilon_0 B} \tag{11.8}$$

This tells us that if, for example, the space-charge density is constant, the velocity increases uniformly across the beam, at a rate inversely proportional to the magnetic field, which is constant. At this stage we cannot assume that the charge density is constant. Let us now add the condition that all the electrons come from a single cathode at a voltage V_c. Then for electrons in the layer at y,

$$\tfrac{1}{2}\, mv^2 = e\,(V_y - V_c) \tag{11.9}$$

Differentiating with respect to y gives

$$mv\,\frac{\partial v}{\partial y} = \frac{e\partial V_y}{\partial y} \tag{11.10}$$

$\dfrac{\partial V_y}{\partial y}$ is, of course, the field E_y, so that

$$v\,\frac{\partial v}{\partial y} = \eta E_y \tag{11.11}$$

Again using Equation 11.6 to replace v, this becomes

$$\frac{\partial v}{\partial y} = \eta B \tag{11.12}$$

Equating this to Equation 11.8 gives the result

$$\frac{\eta\rho}{\epsilon_0} = \eta^2 B^2 \tag{11.13}$$

It follows that the charge density are, in fact, constant, since all the other quantities are constant, and that this quality has been imposed by the use of a single cathode. This type of flow is known as Brillouin flow in crossed fields, and it is seen that it has a similar ideal character to that of Brillouin flow in an axial magnetic field, in which the charge density is also constant throughout the beam. In the crossed-field case, however, the current density must vary across the beam, as a consequence of the velocity variation. The last equation may be written (from Equations 2.7 and 6.6)

$$\omega_p{}^2 = \omega_c{}^2 \tag{11.14}$$

which is also different from the axial-field case.

11.3 Crossed-Field Guns

Beam behaviour and gun design are rather more intractable in crossed fields than in travelling wave tubes. The presence of the fields complicates any experimental beam analysis, and because of beam instabilities, which are discussed later, it was not until the employment of large digital computers became practicable that electron guns with any effective beam convergence could be designed on a theoretical basis. The general arrangement of a gun for an injected

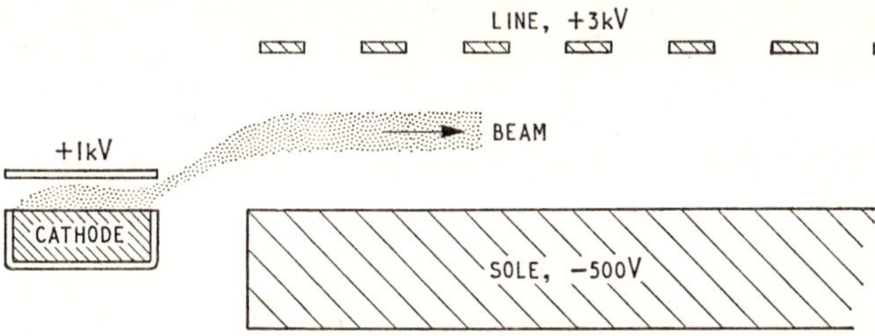

Fig. 11.3. *Crossed-field gun geometry*

beam valve such as a b.w.o. is shown in Fig. 11.3. This geometry was arrived at through experimental investigation.

11.4 Crossed-Field Interaction

In discussing the interaction process, only the case of low charge-density, in which the entire beam has the same velocity, is considered. As is the case with O-type valves we should expect that increasing the beam current, which would increase the space charge, would enable more power to be obtained, although the velocity variation that would be caused might reduce the efficiency of interaction.

The r.f. circuit most widely used in M-type valves is the plane interdigital line shown in Fig. 9.26(a). The interaction mechanism does not, of course, depend on the form of the line; this only affects the characteristics of the device. Some other suitable lines in which the periodic elements lie in a plane are shown, and their characteristics described, in Chapter 9. The r.f. fields produced in a plane through the centre of the beam by a wave propagating along the line have the form shown in Fig. 11.4. The electric field has components in the y and z directions, i.e. in the direction of the d.c.

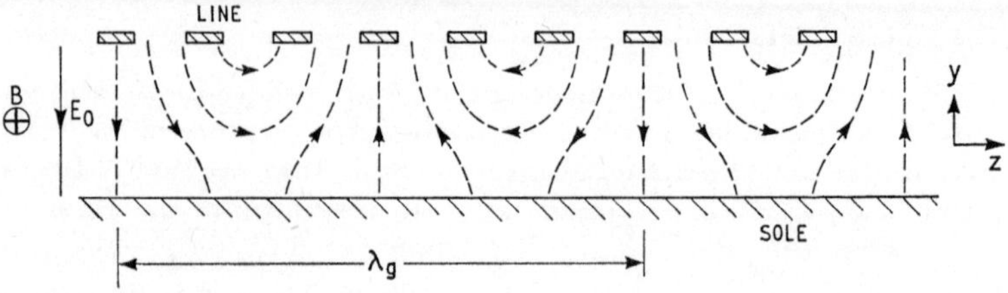

Fig. 11.4. *Form of r.f. electric fields in crossed-field valves*

field E_y and in the direction of motion of the beam. At some points the fields are entirely transverse (E_z=0), while at points half-way between them the field is entirely axial (E_y=0), and the whole pattern reverses every half-wavelength. As in O-type travelling wave devices, cumulative interaction takes place when the velocity of propagation of the wave is the same as that of the electron beam. We consider what happens when a section of the beam lying within one wavelength travels along at the same speed as the r.f. field, which must therefore be made equal to E_o/B, where E_o is the steady electric field. In general the electrons will be executing some form of cycloidal motion, with an angular velocity $\omega_c=\eta B$. When they enter the interaction region there will be added to E_o a smaller r.f. electric field component $E_{\sim}$, which since the r.f. wave is moving at the same speed as the electron will also appear as a steady field. The electron will still execute a cycloidal motion, at an angular velocity ω_c, but will now have a mean velocity of $(E_o+E_{\sim})/B$ which may be slightly altered in direction if the resultant electric field $(E_o+E_{\sim})$ is not in the same direction as the d.c. field. The resultant field is, of course, the vector sum of the two field components, obtained as indicated in the diagram of Fig. 11.5. Figure 11.6 shows the effect of the added field on the electron path for two cases (a) when the electron starts from rest and (b) when the electron is correctly injected at the velocity E_o/B. This diagram shows how the resultant motion is composed of the original motion plus an added cycloidal motion at right-angles to the r.f. field. The period of the added motion is still the same as the period of the initial cycloidal motion, since this depends only on the magnetic field, and the additional r.f. magnetic field is normally negligibly small.

When both the d.c. field and the r.f. field are in the same or opposite directions, as at the points F and G in Fig. 11.7, the effect of the added r.f. field is only to modify the speed of the mean electron slightly. It is made larger when the fields are in the same direction,

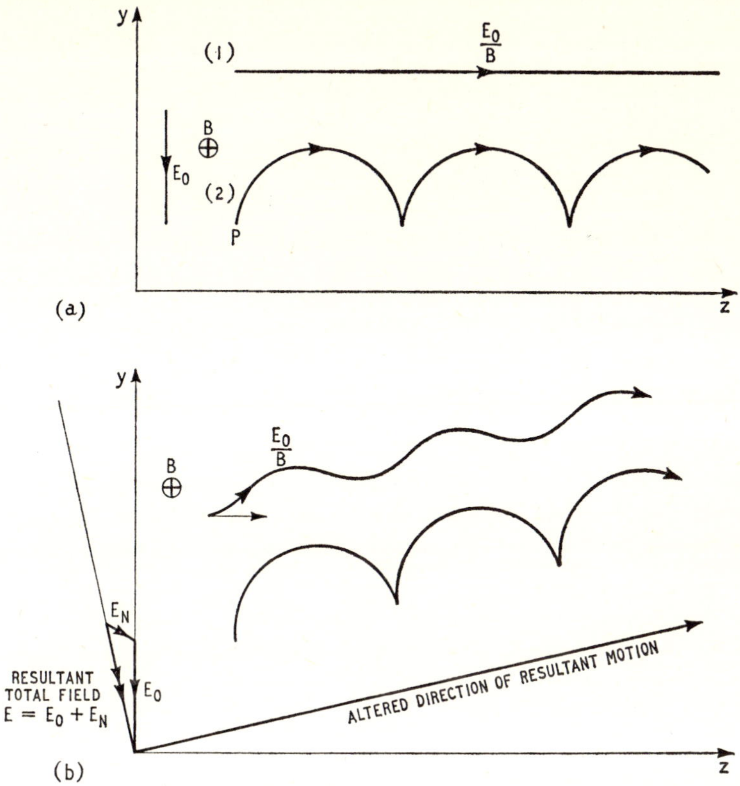

Fig. 11.5. *Change in electron motion due to r.f. fields. (a) Motion of electrons in d.c. fields* (1) *injected at velocity* $V = Eo/B$ (2) *starting from rest at P. (b) Motion of same electrons in presence of r.f. fields*

as at F or slower when the fields are opposing, as at G. The effect of these transverse field components, therefore, is to produce electron bunching into the regions where the motion due to the axial field components is towards the r.f. circuit, which is at a high positive voltage, and out of the regions where the added motion is towards the sole or cathode. The directions of this motion and its effect on the electron beam are shown in Fig. 11.7. Ultimately all the beam would arrive at the r.f. line. Since the electrons in the bunch are moved sideways out of it towards the circuit while other electrons are being accelerated or retarded into the bunch, the bunching does not necessarily involve any increase in charge density.

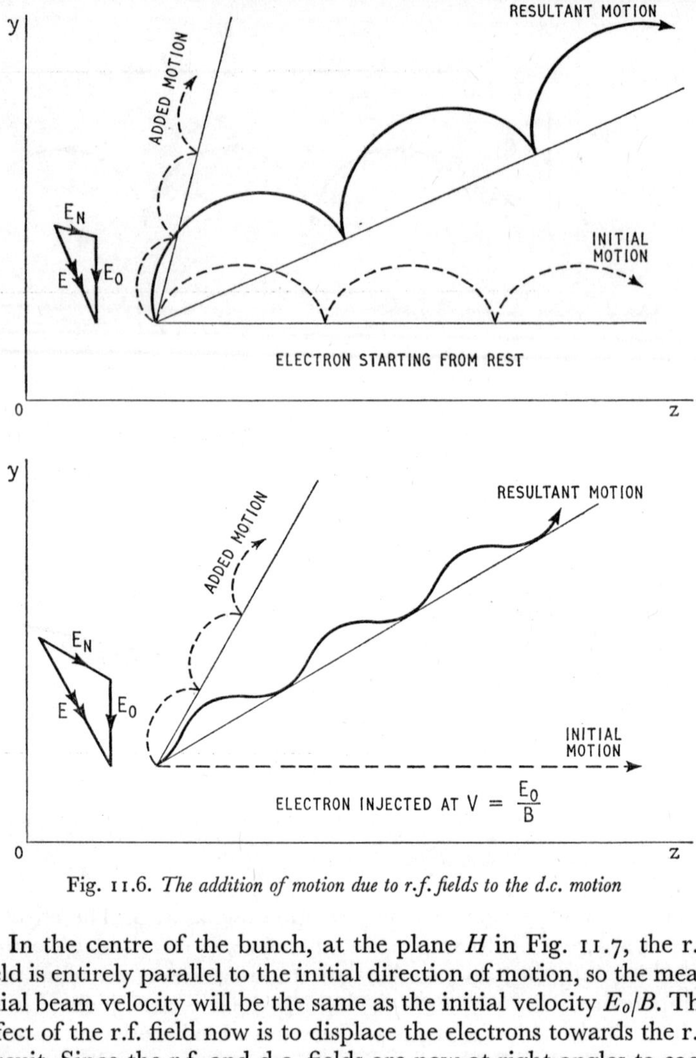

Fig. 11.6. *The addition of motion due to r.f. fields to the d.c. motion*

In the centre of the bunch, at the plane H in Fig. 11.7, the r.f. field is entirely parallel to the initial direction of motion, so the mean axial beam velocity will be the same as the initial velocity E_0/B. The effect of the r.f. field now is to displace the electrons towards the r.f. circuit. Since the r.f. and d.c. fields are now at right-angles to each other, we see that the r.f. field is making the electron move in the direction of the d.c. field force, while the d.c. field makes it move against the direction of the r.f. field force. Its mean transverse velocity is $E_{\sim}/B$, so that in one period of its transverse cycloidal movement it moves a distance

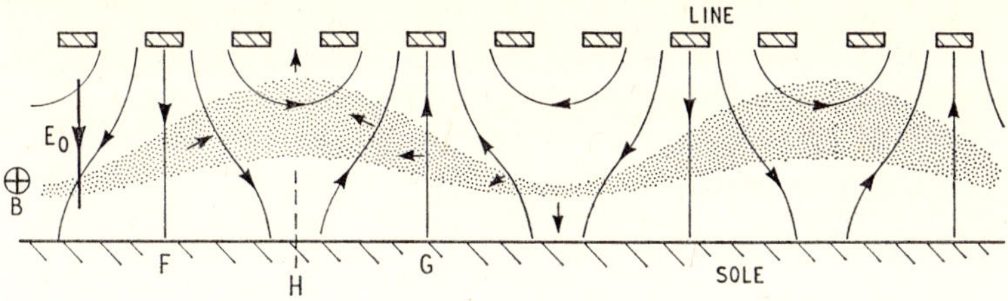

Fig. 11.7. *Crossed-field interaction and beam displacement*

$$\frac{E_{\sim}}{B} \cdot \frac{2\pi}{\omega_c}$$

In doing this, it is moving in the direction of the d.c. field, and so it must absorb an amount of energy equal to force × distance, i.e.

$$\frac{E_0 e E_{\sim} 2\pi}{B\omega_c} \quad \text{or} \quad \frac{2\pi m E_0 E_{\sim}}{B^2}$$

By an exactly similar argument it can be seen that this is also the work done by an electron in moving against the r.f. force, so that each electron is acting as a sort of pivot, which enables energy to be transferred from the d.c. field to the r.f. field. If the electron starts from a cathode at the sole potential, the total energy it has absorbed from the d.c. voltage V_0 in arriving at the line is eV_0. For laminar flow its velocity on arrival is

$$v = \frac{E_0 + E_{\sim}}{B} \tag{11.15}$$

where $E_{\sim}$ is the value of the axial r.f. field close to the circuit. The energy given up to the r.f. field is therefore

$$W = eV_0 - \tfrac{1}{2}m\left(\frac{E_0 + E_{\sim}}{B}\right)^2 \tag{11.16}$$

If the r.f. field is small compared with the d.c. field, which is often the case, this leads to an ideal efficiency

$$\frac{W}{eV_0} = 1 - \frac{V_0}{2\eta d^2 B^2} \tag{11.17}$$

The interaction process is not, as in O-type valves, limited by the ability of the electrons to lose kinetic energy and still remain sufficiently near synchronism to interact; thus very high efficiencies, up to 80 per cent in the case of Amplitrons discussed in the next chapter, can actually be achieved. Another important characteristic of crossed-field valves is that high powers can be achieved without the necessity of very high voltages, hence a given valve can be made much smaller than an O-type valve of comparable power. This is particularly useful at the lower microwave frequencies.

In the electron beam there are velocity modulation and density modulation in the direction of propagation as in an O-type beam; there are also displacement and velocity in the direction of the d.c. field. In the case of a thick beam, even in the ideal Brillouin flow, there is, as has been shown, considerable velocity variation across the beam, and in practice it is doubtful if this flow has been even approximately reached.

11.5 Noise Amplification in Crossed-Field Beams

The simplest beam model amenable to analysis is a thin strip beam, infinitely wide, so that variations across its width to meet the edge boundary conditions can be ignored, and having a finite charge density of σ per unit area. It can be shown that even in such a simple beam, without interaction with any r.f. circuit, growing waves occur, but the analysis is more complex than is appropriate here and and can be found in the references given at the end of the chapter. It follows the usual approach of assuming that waves of angular frequency ω are propagating in the direction of the beam, and have a propagation constant β and hence have a variation in time and distance given by $e^{j(\omega t-\beta z)}$. This is expressed in terms of the d.c. beam quantities, somewhat in the manner used in O-type beam analysis (see Section 6.5.), and yields, for the tangential field components on either side of the flat strip beam,

$$E_1\left[1-\frac{\eta\sigma}{2\,\epsilon_0\omega_c}\left(\frac{\beta}{\omega-v_e\beta}\right)\right]=E_2\left[1+\frac{\eta\sigma}{2\,\epsilon_0\omega_c}\left(\frac{\beta}{\omega-v_e\beta}\right)\right] \quad (11.18)$$

where suffixes 1 and 2 denote regions below and above the beam respectively. A similar equation, but with reversed signs connects the corresponding normal components. Such a beam can only exist in the presence of external conductors to supply the transverse d.c. electric field, and if propagating waves are occurring in the regions between the beam and these conductors their fields at the beam must be those arising in the beam analysis. By equating these it is shown that, for small signals and for waves propagating at near beam velocity,

$$\beta = \beta_e \left(1 \pm j \frac{\eta\sigma}{2\epsilon_0 v_e \omega_c}\right) \tag{11.19}$$

i.e. there is a growing wave, increasing in amplitude, as given by

$$A = A_0 e^{j(\omega t - \beta_e z)} e^{\left(\frac{\beta_e \eta\sigma}{2\varepsilon_0 v_e \omega_c}\right) z} \tag{11.20}$$

11.5.1 Physical Description of Noise Growth

We can see qualitatively why this occurs in the following way. In a uniform well-defined strip beam, as indicated in Fig. 11.8(a) there are equal uniform fields E on either side of it due to the space-charge in addition to the d.c. field, and as long as the beam remains

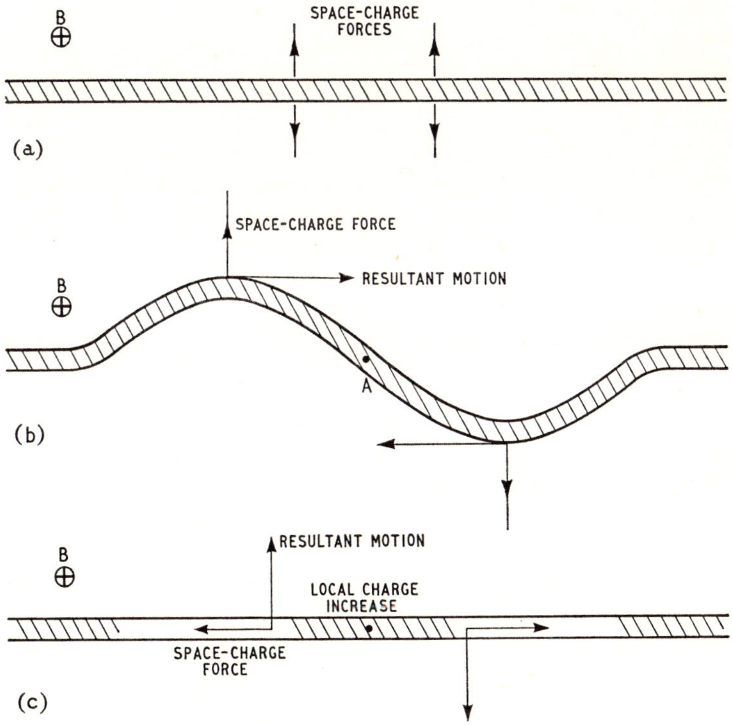

Fig. 11.8. *Instability in thin crossed-field beam*

unperturbed, these space-charge fields produce no forces on the electrons, which continue to travel in a straight line at the speed E/B. Now suppose that in some way some electrons become displaced as is shown in Fig. 11.8(b). Once the electrons have left the beam axis they are affected by the space-charge field acting in the directions as shown. As we have seen, any additional electric field in the crossed-field system produces a mean motion at right-angles to itself, so that the resultant motion of these electrons will be as shown. The effect of this will be to cause an increase in the charge density, which will be centred around the point A. Now consider the effect of this local increase of charge occurring in the beam, as shown in Fig. 11.8(c). This will produce local axial space-charge fields as shown, which will in turn lead to mean motion in the directions of the arrows. This motion will cause a displacement of the beam of just the kind considered above. This situation, in which one event causes another, which then causes an increase of the first, is clearly an unstable one and so it would be expected that any such perturbation on the beam will increase as the beam progresses.

Fluctuations in the beam are always present, arising from fluctuations of density and velocity of emission from the cathode, and as these become amplified, even before the beam enters the circuit interaction region, they produce an increasing amount of noise on the beam. This has been one of the most serious disadvantages of crossed-field interaction; its effect is to limit the amount of gain that can be usefully employed to about 15 to 20 dB. If it is attempted to use greater gain, the lower level signal required at the input deteriorates due to noise. For this reason consideration of crossed-field amplifiers has usually been as low-gain, high-power, high-efficiency amplifiers in the output stages of high-power radar systems.

11.6 Crossed-Field Backward-Wave Oscillators

By far the most common example of the type of valve dealt with above is *the crossed-field backward-wave oscillator* (*M-Carcinotron*). These have been made over most of the microwave spectrum, and except at the highest frequencies, the actual valve arrangement has been roughly constant, although the magnetic circuit and encasement have shown some variations. The r.f. structure used is an interdigital line, usually made from two copper combs, as shown in Plate 11.1, which are interplaced and fixed to a solid copper component. This makes the ω-β characteristic curved at the lower frequency end, since a low-frequency cut-off now occurs. The tuning curve of a 3 Gc/s valve is shown in Fig. 11.9. The frequency of a backward-wave oscillator is tuned by varying the beam velocity, which in the case of a crossed-field valve is given by E/B. The d.c. field E is fixed by the voltage applied across the interaction gap, so that the velocity varies directly as this voltage. In this respect the valve differs from the low-power O-type valves, in which the velocity varies as the square root of the voltage.

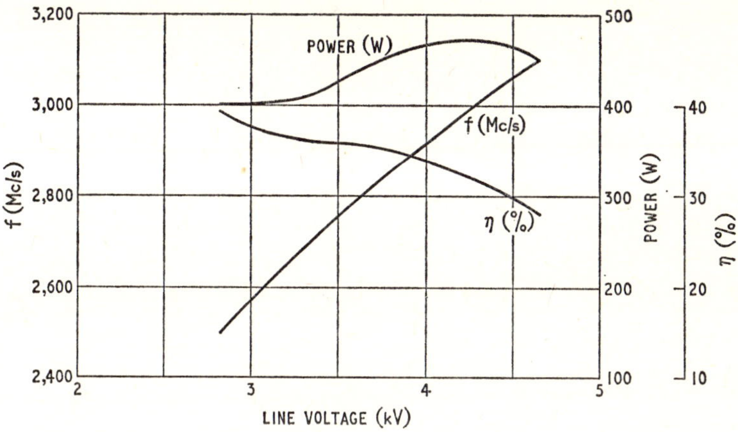

Fig. 11.9. *Typical performance of* 3,000 Mc/s *crossed-field backward-wave oscillator at* 350 mA *line current*

The general arrangement and construction are shown in Fig. 11.10. The structure is bent into a circular form so that the magnetic field may be more easily provided by a permanent magnet. This does not affect the interaction process. The r.f. line is normally fixed around the inside of the body, which is earthed, and the output coupling in the example shown, which is a valve operating at 10 cm, is made by extending the end finger of the line so that it becomes the inner of a coaxial line. At higher frequencies a waveguide coupling is used. Great care has to be taken in obtaining a good match both from the line to the external load, and at the termination at the far end of the line, in order to obtain an output power that does not vary rapidly with frequency. If either of these matches is inadequate, the tuning curve of frequency against voltage becomes uneven and there may even be frequency 'holes', i.e. breaks in the frequency range in which no output can be obtained. The d.c.

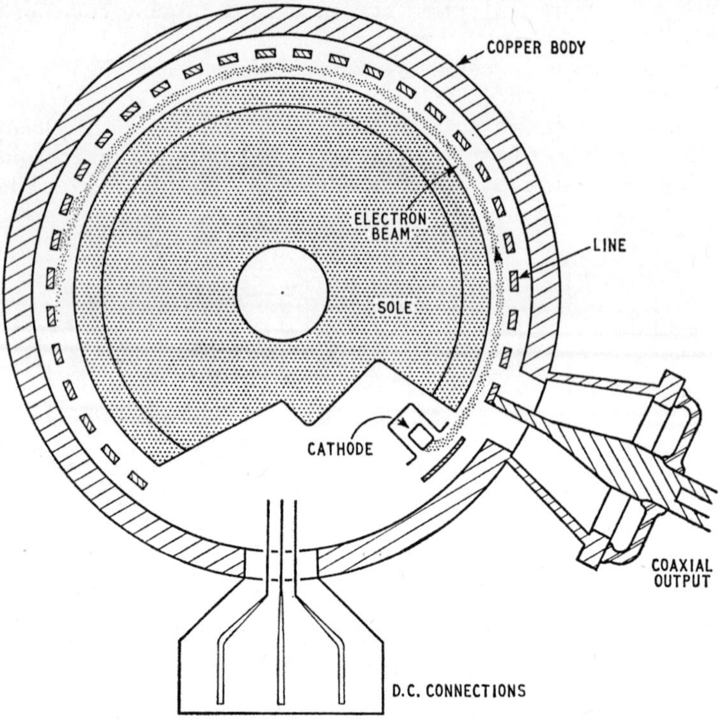

Fig. 11.10. *General arrangement of* 3,000 Mc/s *crossed-field backward-wave oscillator*

transverse field is produced by a large negative voltage of a few kilovolts applied to the sole, which is supported as shown from a glass or ceramic coaxial vacuum seal from the body. It can be seen that the gap between the sole and the line, across which the electric field is developed, is very small, and any variations in this will produce variations in the electric field and hence in the electron velocity. The sole must therefore be placed and maintained in position with great accuracy. This is made more difficult by the fact that it may receive considerable electron bombardment when the valve is in use and at the same time is rather difficult to cool, while the r.f. line itself is also heated by the electron beam, and hence moves relative to the sole. These difficulties—as well as those of making the r.f. lines—increase with frequency, and have so far set a limit of about 30 Gc/s to the upper frequency of this type of valve.

The magnetic field is obtained from external permanent magnets which may be of the horseshoe type, as shown in Plate 11.2, or an enclosing symmetrical shape.

The c.w. power obtainable from cross-field b.w.o.'s lies between a few kilowatts at 500 Mc/s and 10 W at 40 Gc/s. The variation across the tuning range is typically as indicated in Fig. 11.9, which shows the power-frequency curve of the valve shown in Plate 11.2.

11.7 Injected Beam Amplifiers

More rarely met is the *injected-beam crossed-field amplifier*, as comparatively few valves of this type have appeared, mostly due to the difficulties with gain and stability referred to earlier. The interaction principle is the same as that in the backward-wave oscillator, but now takes place with a forward wave, i.e. one in which the phase and group velocities have the same direction. A suitable circuit, for example, is the meander-line.

11.8 Continuous-Cathode Valves

Linear continuous-cathode valves, either oscillators or amplifiers, have met with very little success, although interest has continued in them for well over ten years, and has recently been increased because it is expected that such valves would have less variation of output phase with amplitude than O-type amplifiers. In this arrangement all the electrons start from rest at the continuous cathode surface, and since this surface is very much greater than is possible in an injected beam valve, much greater currents and therefore powers are possible. The chief difficulty with the arrangement is the familiar one in crossed-field systems, of instabilities in the beam itself, which both interfere with the desired signal interaction and upset the required electron flow. From a d.c. model without space charge it would be expected that, as usual, the mean electron velocity would be E/B, while the maximum instantaneous velocity at the limit of the cycloid would be $2\,E/B$ (Fig. 11.5), since, taking an origin in the cathode, the path of an electron would be given by

$$y = \frac{E}{\omega_c B}(1 - \cos \omega_c t) \qquad (11.21)$$

$$z = \frac{E}{B}\left(t + \frac{1}{\omega_c}\sin \omega_c t\right) \qquad (11.22)$$

and the velocities in the y and z directions would be

$$\frac{\mathrm{d}y}{\mathrm{d}t} = \frac{E}{B} \sin \omega_c t \tag{11.23}$$

$$\frac{\mathrm{d}z}{\mathrm{d}t} = \frac{E}{B} (1 - \cos \omega_c t) \tag{11.24}$$

The maximum distance that an electron can go in the direction of the field is

$$y_{max} = \frac{2E}{\omega_c B}$$

so that no current should flow to the line unless $\frac{2E}{\omega_c B}$ is greater than d, which may be rewritten as

$$V_o > \frac{\eta B^2 d^2}{2} \tag{11.25}$$

This gives a cut-off voltage, which depends on B and d, below which line current should not occur, and which corresponds to the magnetron cut-off voltage described in the next chapter. This is what should happen when space charge is negligible. When space-charge effects are large it might be hoped that the electrons emitted from the cathode would go into the Brillouin flow described earlier in this chapter, since the boundary condition of $v_z = 0$ at the zero equipotential (i.e. the cathode) is being met. In fact, the growth instability already described, and the multi-velocity nature of the beam, cause large disturbances, and it is very rarely that any cut-off can be obtained. Large noise-like outputs occur, and it is only when a very large r.f. signal is applied, which effectively takes charge of the beam, that the arrangement works as it should.

It is hoped that the foregoing may have given some idea of the intractability of the problems and the difficulty of analysis in crossed-field systems when space charge is present. Up to the time of writing only the injected-beam backward-wave oscillator has been made successfully in any numbers, although many partially successful experimental amplifiers have been made, and at present, as more or less continuously for the last ten years, success seems only just around the corner.

BIBLIOGRAPHY

OKRESS, E., *Crossed-Field Microwave Devices, Vols. I and II*, Academic Press (1961).

Chapter 12

CYLINDRICAL CROSSED-FIELD VALVES, THE AMPLITRON AND MAGNETRON

Introduction

The *Amplitron* and *magnetron* are in conception and construction similar to continuous injection crossed-field valves arranged in a cylindrical shape, but with the extremely important difference that the electron beam can circulate completely round the cylindrical region between the cathode and the line as shown in Fig. 11.1(d). The use of this arrangement results in d.c. characteristics for these valves quite different from those in the last chapter, and also leads to the production of very high efficiencies. Before considering the construction or operation of the valves, it is necessary to study briefly the electron flow which occurs.

12.1 The Smooth-Anode Magnetron

If the arrangement of Fig. 12.1(a) is considered, in which a cylindrical cathode of radius r_c is along the axis of a smooth cylindrical anode of radius r_a, with a magnetic field B also parallel to the axis, and a positive voltage V_0 relative to the cathode is applied to the anode, electrons emitted from the cathode are initially accelerated towards the anode by the field E_0, but as soon as they acquire any velocity they are deflected into roughly cycloidal paths by the action of the magnetic field. If the magnetic field is large enough,

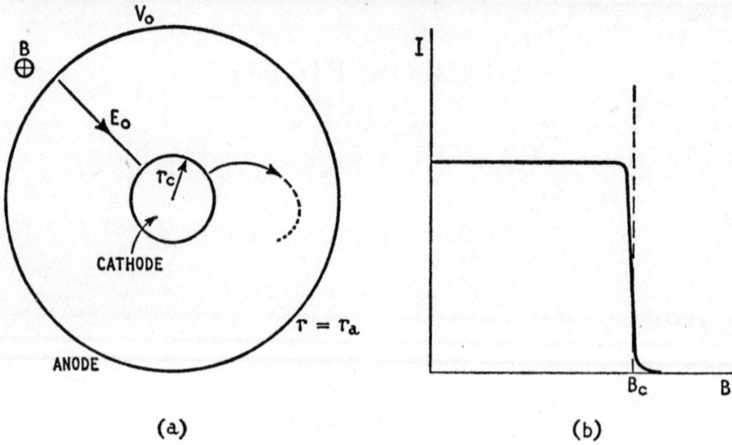

Fig. 12.1. *Magnetron cut-off: (a) cylindrical magnetron (b) cut-off characteristic*

the deflection is sufficient to prevent the electrons ever reaching the anode, and they are returned to the cathode. For any applied voltage, there is a critical magnetic field, called the cut-off field, above which anode current cannot flow, as indicated in Fig. 12.1(b). This behaviour, which has been known since 1912, gave rise to the word magnetron considerably before the use of the arrangement in microwave valves.

12.2 Magnetron Cut-Off

The cut-off effect in this simple model, in which space-charge effects are neglected, can be readily obtained. If the potential at radius r is V_r, then the kinetic energy of an electron emitted with zero velocity from the cathode (at $V=0$) is given by

$$e\, V_r = \tfrac{1}{2}\, m\, (v_r^2 + v_\theta^2) \qquad (12.1)$$

where v_r is the radial velocity and v_θ is the azimuthal velocity. It can easily be shown by an electrostatic argument that E_r varies as $1/r$ and hence that

$$V_r = V_o \frac{\log_e \dfrac{r}{r_c}}{\log_e \dfrac{r_a}{r_c}} \qquad (12.2)$$

The maximum radius reached by an electron will be at the point when its radial velocity v_r becomes zero; hence at this radius

$$v_\theta^2 = 2\eta V_o \frac{\log_e \dfrac{r}{r_c}}{\log_e \dfrac{r_a}{r_c}} \tag{12.3}$$

The azimuthal velocity can also be obtained by equating the moment of the deflecting force to the rate of change of angular momentum, i.e.

$$rBe \frac{dr}{dt} = m \frac{d}{dt}(rv_\theta)$$

which can be rewritten

$$\eta Br dr = d(rv_\theta)$$

which by integration gives

$$\frac{\eta B}{2}(r^2 - r_c^2) = rv_\theta$$

or

$$v_\theta = \frac{\eta B}{2r}(r^2 - r_c^2) \tag{12.4}$$

This equation gives the azimuthal velocity at all points in the anode-cathode space, whereas Equation 12.3 only gives it at the extreme of the electron cloud, where $v_r = 0$. At this value of r, therefore,

$$\frac{\eta^2 B^2}{4r^2}(r^2 - r_c^2)^2 = 2\,\eta\,V_o \frac{\log_e \dfrac{r}{r_c}}{\log_e \dfrac{r_a}{r_c}} \tag{12.5}$$

When the electron flow is just cut-off, this condition holds just at $r = r_a$, when Equation 12.5 simplifies to

$$V_c = \frac{\eta B^2}{8} r_a^2 \left(1 - \frac{r_c^2}{r_a^2}\right)^2 \tag{12.6}$$

This gives the well-known cut-off parabola, Fig. 12.2, relating the cut-off voltage V_c to the field for a given geometry, and we may note that even when d.c. space-charge forces are taken into account the equations used are still true at the anode, where V is always V_o, the applied voltage, so that the same cut-off curve is obtained, although the voltage variation across the cathode-anode space will be different.

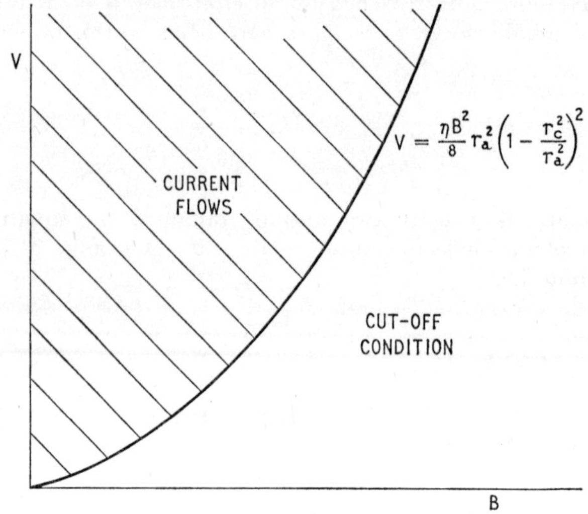

Fig. 12.2. *Magnetron cut-off parabola*

12.3 Noise Generation in Magnetron Current Flow

Inside the maximum radius given by Equation 12.5 there is a cloud of electrons with varying and intersecting trajectories, but in general with increasing azimuthal velocities as the maximum radius is approached. This is still true when space-charge effects are taken into account, and as long as there is no variation of the cloud around the space, the azimuthal velocity corresponding to a particular radius will be given by Equation 12.5. For the purpose of understanding Amplitron and magnetron behaviour, it is convenient therefore to regard the electron cloud as a series of shells rotating with increasing angular velocity as we go from the cathode to the anode.

It has been seen in the previous chapter that in a linear system an electron stream such as is comprised by one of these shells will, since it is in a crossed-field system, produce amplification along its length. Now that the beam is bent into a circle and is continuous, this amplification process will produce growing standing waves at any frequencies where the total distance around the stream is an integral number of wavelengths. Since there are many such streams, all having different angular velocities, there will be many such frequencies, and so the electron cloud will produce large quantities of noise over a wide range of frequencies. This effect is a large signal one, i.e. the r.f. perturbations are comparable with the d.c. quantities of the beam, and the d.c. nature of the beam is quite destroyed by their existence. It is for this reason that it is not profitable to attempt here any detailed study of d.c. electron flow in which space charge is taken into account, such as Brillouin flow, since due to the r.f. instability it is unlikely that any such flow can have, at best, anything more than a very transitory existence.

For example, in a magnetron diode of the sort we have been considering, with an anode circumference of 8 cm, and running in a field giving a cut-off voltage of 10 kV, the azimuthal velocity of the outer electrons as the cut-off voltage is applied is about 0·2**c** and this is roughly the velocity of the growing waves. A wave at 3,000 Mc/s ($\lambda_0 = 10$ cm) would therefore have a wavelength of 2 cm around the stream, and would give a resonant pattern of 4 wavelengths ($n=4$). Similarly, patterns at 3,750 Mc/s ($n=5$), 4,500 Mc/s ($n=6$), etc. could also occur. Clear patterns of this sort might be expected if there were only a thin strip of electrons near the anode, but the presence of all the other electrons at lower velocities and smaller radii gives infinite sets of other possibilities. If a probe is inserted into such a system a high-level noise output is obtained over a very wide range of frequencies, with spikes at particularly favourable frequencies. The presence of all this r.f. power does, as stated, considerably alter the electron-cloud characteristics, and the ideal cut-off curve of Fig. 12.1(b) can be obtained only when the cathode emission is severely temperature limited.

When a r.f. circuit is present, so that r.f. fields can propagate around the anode, the picture is even more changed. A mathematical analysis of the behaviour in the presence of small r.f. fields shows that anode current starts to flow at a threshold voltage considerably below that given by the cut-off parabola, but we shall be returning to this point later in the chapter.

12.4 The Amplitron

The magnetron diode considered above has a cylindrical cathode along the axis of a smooth cylindrical anode. In the Amplitron this geometrical arrangement is used, but in place of a smooth cylindrical anode there is a r.f. slow-wave circuit as shown diagrammatically in Fig. 12.3. It is to be noted particularly that, unlike the electron beam, the r.f. line is not continuous right round the anode plane. It has r.f. input and output couplings at its ends, and the short region of

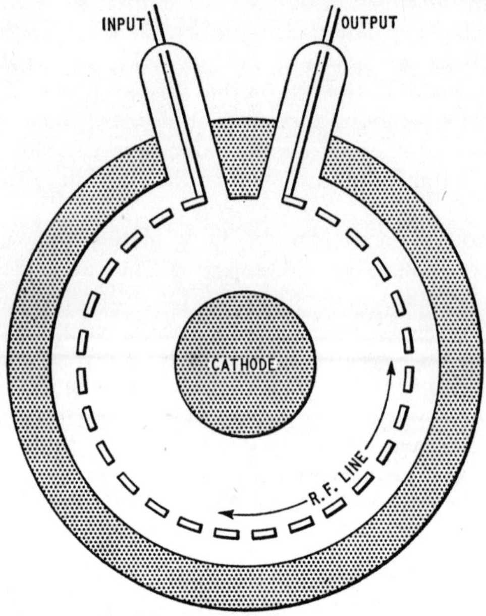

Fig. 12.3. *Arrangement of amplitron*

the anode between these has a smooth surface and hence does not propagate the slow wave. If no r.f. power is applied to the input coupling, the circuit will pick up signals from the noise of the electron cloud and propagate those that lie within its pass-band. Some frequencies may be propagated at a velocity which roughly matches their velocity in the electron cloud, and these will be further amplified. The result will be a large noise-like output from the output coupling. Suppose now that a gradually increasing signal is applied to the input coupling. At first this will be insufficient to modify the large r.f. disturbances already existing on the beam, but eventually, when the r.f. fields of the applied signal become comparable with those already existing, more of the beam energy will be directed into r.f. bunching at the applied frequency. As soon as this starts to happen the re-entrant nature of the beam will cause bunching at this frequency to build up very rapidly, and as it does so the magnitude of the disturbances at all other frequencies will rapidly be suppressed. At the same time the d.c. conditions of the valve will adjust themselves suitably. Very strong interaction will then occur at the frequency of the applied signal, and high-power output can be obtained from the output coupling. The efficiency under these conditions can be as high as 80 per cent. The input-output characteristics of an Amplitron are therefore as shown in Fig. 12.4. Over a large region, until the input signal is large enough

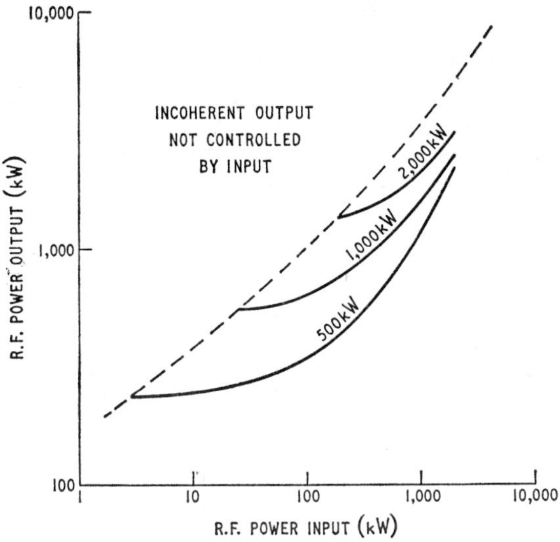

Fig. 12.4. *Amplitron characteristics for different modulator inputs* (*after* COLLINS, 1948)

to capture the space-charge cloud, the output is incoherent, and not dependent on the applied signal. Over the right-hand side of the diagram there is a high-power output at the input frequency, showing a power gain of between 2 and 10 db.

The electron bunching produced in the space charge by the interaction with the wave on the circuit is, of course, carried round with the beam across the region between the output and input couplings. At some frequencies these bunches will arrive at the input to be in phase with the bunching forces there, and so the beam will be effectively prebunched. At these frequencies therefore, the gain and efficiency will be increased. On the other hand, there will be some frequencies at which the re-entering bunches are in antiphase to those being formed, giving degradation of gain and efficiency. Generally, if the phase of the re-entering bunch is within about 45° of the phase of the incoming wave, trapping occurs and the gain is not greatly impaired. This limits the bandwidth of the amplifier to about 10 per cent. The frequencies at which these respective effects recur will depend on the gap between input and output couplings, the operating voltage and magnetic field, and the space-charge density, among other things. They may also be affected by the power supply impedance, as discussed later.

12.5 Amplitron Applications

From the foregoing it will be apparent that the logical use of an Amplitron is as a low-gain but highly efficient output stage, and most Amplitrons have been designed for very high-power pulsed operation for radar use, when they are driven by either a lower-power magnetron or a t.w.t. amplifier chain. Some examples of currently available Amplitrons and their performances are shown in Plate 12.1.

It should be mentioned that Amplitrons were invented and are solely made by the Raytheon Mfg. Co. and hence the relevant information and illustrations have come directly or indirectly from that company.

12.6 Interaction with Power Supply

Amplitrons and magnetrons are most frequently used as high-power pulsed valves, and these behave rather differently from other valves in the way that they react with their power supplies. A high-power pulsed supply, known as a pulse modulator. is arranged to deliver a large energy pulse at a constant voltage into a load which is determined by the desired voltage and current of the magnetron. One commonly used method employs the discharge of an artificial transmission line, i.e. a combination of inductances and condensers as shown in Fig. 12.5. The characteristic impedance of a real lossless transmission line is given by $Z=(L/C)^{\frac{1}{2}}$ where L is the inductance per unit length and C is the capacitance per unit length, and similarly the characteristic impedance of the artificial line is $(L'/C')^{\frac{1}{2}}$. If such a line is charged to a high voltage, V, and then suddenly connected by a fast-acting low-impedance switch (such as a thyratron) to a load equal to its characteristic impedance, it delivers its stored energy into the load in the form of a short uniform pulse, during which the voltage $V/2$ across the load and the current flowing through it remain constant. The length of the pulse for a line of n sections is $n(LC)^{\frac{1}{2}}$, and is commonly of the order of a microsecond.

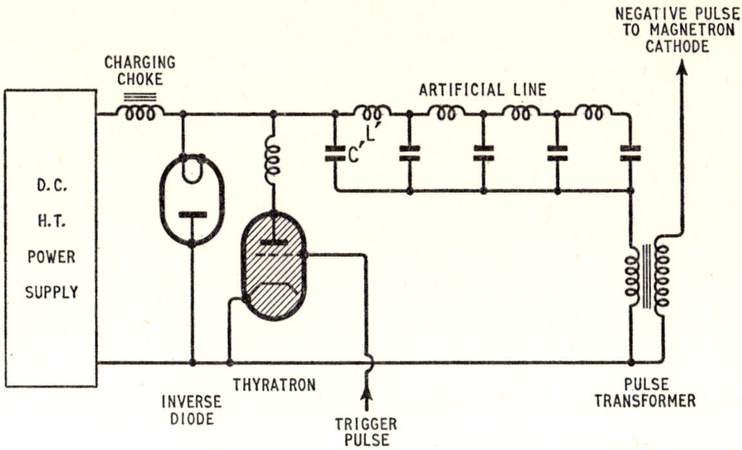

Fig. 12.5. *Pulse modulator*

It is just the time that an electric wave takes to travel to the far end of the line and back again. It is usually inconvenient to have an artificial line with a characteristic impedance of more than a few ohms, while typical magnetron impedances are about 1,000 Ω. The necessary impedance matching is achieved by means of a pulse transformer, which is fundamentally similar to an ordinary transformer, but which has its self- and leakage-capacitance and inductance carefully arranged so that it will pass on the applied pulse faithfully to the load. The circuit of such a system is shown diagrammatically in Fig. 12.5. Now, this modulator will only produce the designed pulse when connected to the designed load. If connected to an open circuit, the pulse is distorted as shown in Fig. 12.6 and the voltage rises to a much higher value, while if the load impedance is reduced the secondary circuit of the pulse transformer is underdamped, and ringing occurs. As has been seen, a magnetron type of valve does not pass any current until its threshold voltage is reached, so that while the pulse is building up the valve looks like an open circuit. When the threshold voltage is reached the valve starts to conduct, and the current rises very rapidly, holding the voltage to the operating value, while at the end of the pulse, as soon as the pulse voltage falls below the threshold value, current ceases and the valve again resembles an open circuit. The valve thus behaves to a large extent as a constant voltage device, this voltage being determined by the valve geometry and the magnetic field, while the

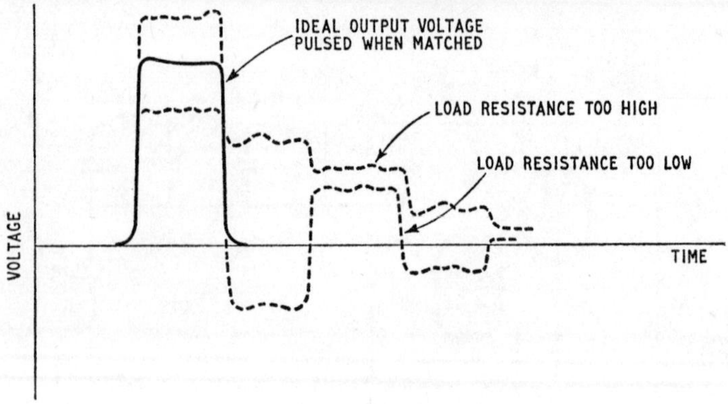

Fig. 12.6. *Modulator output waveform*

current through it varies in such a way as to keep this voltage constant. It will be seen that in a valve with large r.f. powers present the operating voltage is altered by the presence of the r.f. and the exact voltage depends on the frequency. This characteristic makes valves of this type relatively easy to use. As long as a sufficiently powerful pulse is supplied which is capable of reaching at least the operating voltage, the exact voltage will be determined by the valve, and variations in the pulse current will be reflected in output power fluctuations. This behaviour is in contrast to that of high-power pulsed klystrons or travelling wave tubes, where the valve impedance is much more nearly constant and the pulse modulator must be much more critically designed so that the output voltage is constant at exactly the required value throughout the pulse.

12.7 The Multiple-Cavity Magnetron Circuit

The magnetron is, historically, the oldest microwave valve and paradoxically perhaps the least completely understood. High-frequency oscillations in valves using the magnetron geometry have been studied since about 1920, and the existence of travelling wave interaction in a magnetron with multiple segments in the anode was recognised in 1935. However, the road by which the approach to understanding was made historically is an unnecessarily difficult one, and it is perhaps easier to attempt this using the concepts that we have attempted to establish in earlier chapters. Some of the behaviour of the electron beam in a smooth circuit has already been dealt with. The types of circuit actually used are shown in Fig. 12.7.

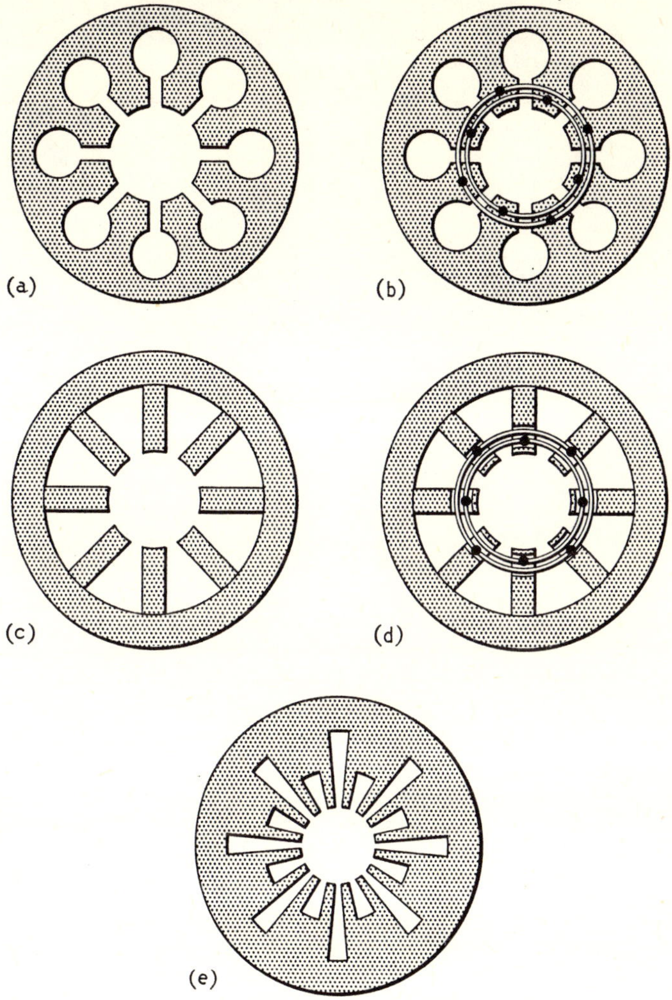

Fig. 12.7. *Some magnetron circuits: (a) unstrapped cavity (b) strapped cavity (c) unstrapped vane-block (d) strapped vane-block (e) rising-sun block*

All these consist of a number of cavities arranged round the anode block. To understand their r.f. behaviour, let us suppose that such a system is opened out into a straight system, as in Fig. 12.8(a).We should expect such a structure to propagate waves over a narrow

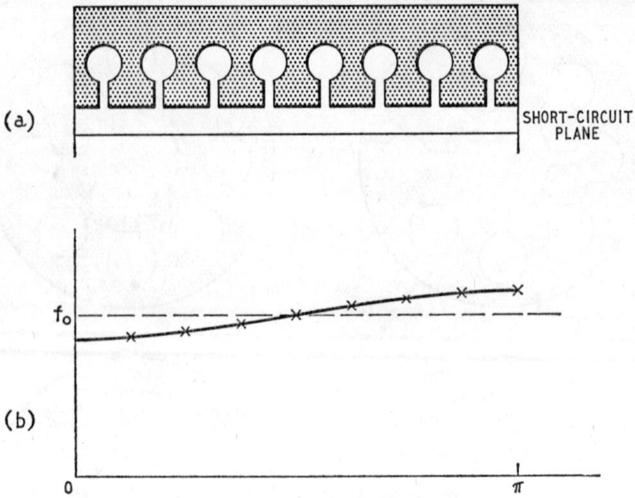

Fig. 12.8. *Propagation of coupled-cavity line: (a) linear model (b) phase change per cavity*

frequency band, centred on the resonant frequency of the cavity, giving a dispersion diagram of the type shown in Fig. 12.8(b). It will be recalled from Chapter 3 that such a curve can be obtained by terminating a finite length of structure by short circuits in two planes of symmetry, and measuring the frequencies at which resonance can be obtained. These are when the system is an integral number of half-wavelengths long. An eight-section length of line would therefore give eight such resonances, when its electrical length would be $\lambda/2$, λ, $3\lambda/2$, 2λ, $5\lambda/2$, 3λ, $7\lambda/2$, and 4λ respectively, giving

$$\frac{\pi}{8}, \frac{\pi}{4}, \frac{3\pi}{8}, \frac{\pi}{2}, \frac{5\pi}{8}, \frac{3\pi}{4}, \frac{7\pi}{8} \text{ and } \pi \text{ rad}$$

for the corresponding phase changes per cavity. Such a line thus has eight *modes* of oscillation, which may be denoted by the numbers of half-wavelengths along the line as $m=1$, $m=2$, etc. Now consider how this picture is changed by reverting to the cylindrical form from which we started. The propagation of the line, we may suppose, will not be changed greatly by the change from the curved to the straight form, but because the short-circuiting planes are removed oscillations will now be observed only when the electrical distance around the block is a whole number of wavelengths, so that a wave progressing around the block reaches its point of starting at the same phase in which it started. Only the resonances corresponding to even numbers of *m* survive, and such a block would therefore have four resonances, denoted by $n=1$, $n=2$, $n=3$ and $n=4$, where n is the number of wavelengths around the block. Of these, the most important is that giving π phase change per cavity (in this case $n=4$) since it is in this mode, called the π-mode, that magnetrons are always designed to oscillate. In this mode, the voltages on all the alternate segments are rising and falling in phase together, and the state of charge, current and electric and magnetic field distribution throughout one cycle are shown diagrammatically in Fig. 12.9. The earliest

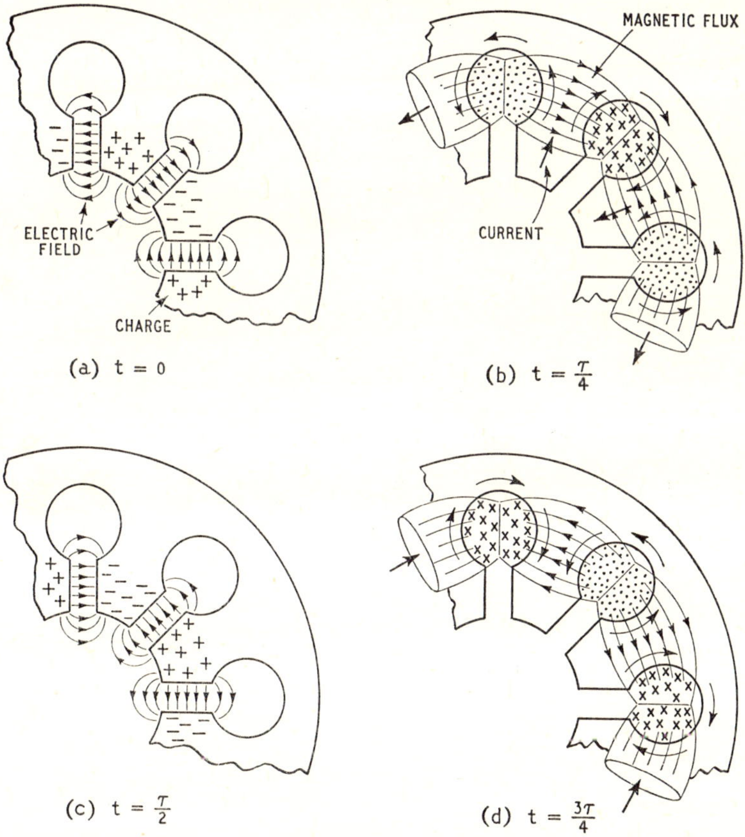

Fig. 12.9. *π-mode oscillation in a magnetron block*

high-power copper-block magnetrons had a simple anode structure just like that of Fig. 12.7(a), although, as will be seen, this is usually modified by strapping as shown in Fig. 12.7(b), in modern valves.

12.8 Minimum Voltage of Operation

For a magnetron block at the frequency of its n-mode the wavelength measured at the anode surface is

$$\lambda = \frac{2\pi r_a}{n}$$

The azimuthal velocity v of the wave is therefore

$$v = \frac{\omega r_a}{n} \tag{12.7}$$

To match the electron velocity at this radius to the wave velocity, we use Equation 12.4.

$$v_\theta = \frac{\eta B}{2r_a}(r_a^2 - r_c^2)$$

giving

$$B = \frac{2\omega r_a^2}{\eta n\,(r_a^2 - r_c^2)} = \frac{2\omega}{\eta n\left(1 - \dfrac{r_c^2}{r_a^2}\right)} \tag{12.8}$$

This gives the minimum field required for operation at a given n and frequency, which field thus increases as the frequency increases. Very high-frequency magnetrons therefore require high magnetic fields.

Now if oscillation takes place just at the cut-off condition, the cut-off equation (12.6), i.e.

$$V_c = \frac{\eta B r_a^2}{8}\left(1 - \frac{r_c^2}{r_a^2}\right)$$

applies at the same time as Equation 12.8, and together these give

$$V = \frac{B r_a^2 \omega}{4n} \tag{12.9}$$

12.9 Threshold Voltage

This gives values of voltage lying on the cut-off curve at which oscillation in the n-mode can be expected to start. However, magnetrons are never run under these conditions, since the electronic efficiency would be extremely low. In general a magnetic field considerably higher than the minimum field is chosen, since this gives high efficiency. With this type of operation it can readily be shown that the threshold voltage (i.e. the voltage at which anode current will commence to flow for vanishingly small r.f. voltage on the segments) is given by

$$V = \frac{Br_a{}^2\omega}{2n}\left(1 - \frac{r_c{}^2}{r_a{}^2}\right) - \frac{r_a{}^2\omega^2}{2\eta n^2} \qquad (12.10)$$

This equation is a linear relation, and therefore gives, as shown in Fig. 12.10, a series of straight lines starting from the cut-off curve. The higher the value of n, the more the threshold line lies below the cut-off parabola. Usually for practical high power magnetrons the

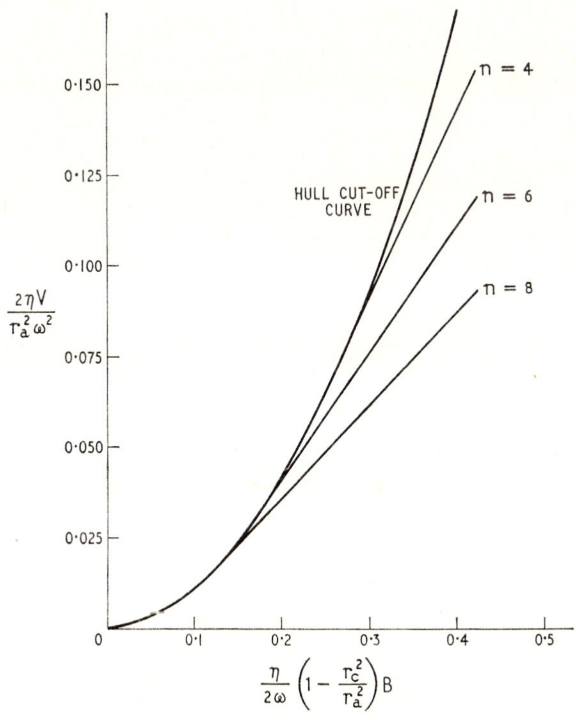

Fig. 12.10. *Magnetron oscillation threshold-voltages*

ratio $\dfrac{V_{threshold}}{V_{cut\text{-}off}}$ is about 0·5, which gives a theoretical electronic efficiency of about 80 per cent.

This is a simple theory, but it is found to be quite accurate in practice in predicting the way in which the lowest voltage at which oscillation will start varies with the magnetic field, in a valve designed to operate with a given n and ω.

12.10 Operation in the π-Mode

Since the mode with the highest n-number is the π-mode, this has the lowest operating voltage, and hence is the first mode in which oscillations might be expected as the voltage is raised. In an unstrapped block, however, the frequencies of all the modes are comparable, and this behaviour becomes complicated by the loading of the desired mode by the coupling circuit, and by the amount of current available from the cathode. Either of these, or other mechanisms like coupling through the end-space or cathode resonances, can cause other modes to be excited, giving a 'mode-change' performance. When this happens the magnetron may change its frequency and operating voltage during a pulse, or from pulse to pulse, generally causing malfunction of the transmitter. This problem of mode stability is and has always been the main problem in magnetron design and development; how to make a valve oscillate and keep on oscillating in its π-mode. An enormous amount of thought and ingenuity has been applied to this problem and every type of magnetron that can be bought represents a particular solution that has been found.

12.10.1 Effect of Coupling

Consider first the effect of the coupling. A simple waveguide transformer coupling, as shown in Fig. 12.11, produces a loading at one point in the anode block. Now the π-mode is symmetrical around the block about any cavity, but for the lower n-number modes this is no longer true, and they can orient themselves with respect to the coupling so that they are lightly or heavily loaded. This means that for every mode other than the π-mode, there are two possible frequencies, very close together, the lower being heavily coupled to the external waveguide and the upper being lightly coupled. They are therefore said to be *degenerate*. A lightly loaded resonance can always be more easily excited than a heavily loaded one, so one set of possibilities of mode changing exists here.

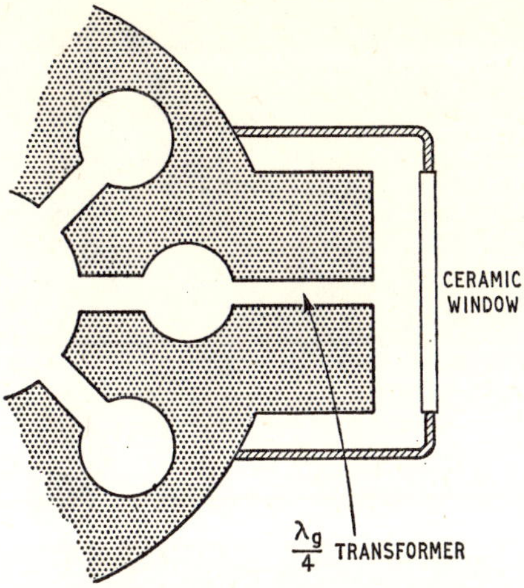

Fig. 12.11. *Waveguide transformer coupling*

12.10.2 Effect of Inadequate Emission

The second reason listed for mode changing is inadequate cathode emission. When a magnetron is oscillating at high power in the π-mode, the space charge is deformed by the r.f. fields into a spoke-like form as shown in Fig. 12.12. As the power input to the modulator is increased, the magnetron tries to pass more and more current to hold the voltage down to that of the π-mode, resulting in more and more r.f. output, but eventually the limit is reached of the available emission from the cathode. When this happens the valve usually jumps into the next lower mode, often called the '$\pi-1$' mode, since its n number is $n_{\pi}-1$. This has one less spoke of space charge than the π-mode, and in addition has a higher operating voltage, both of which factors permit a reduction in the total current. Of course, in a well-designed magnetron this does not happen in the operating range, but if a valve is approaching the end of its cathode life this may frequently be revealed by mode-changing behaviour.

12.11 Space-Harmonics

Further possibilities of mode changing arise through interaction with space harmonics of any given mode. The existence and significance of space harmonics in periodic structures has already been discussed in Chapter 9, but it was in the study of magnetron circuits

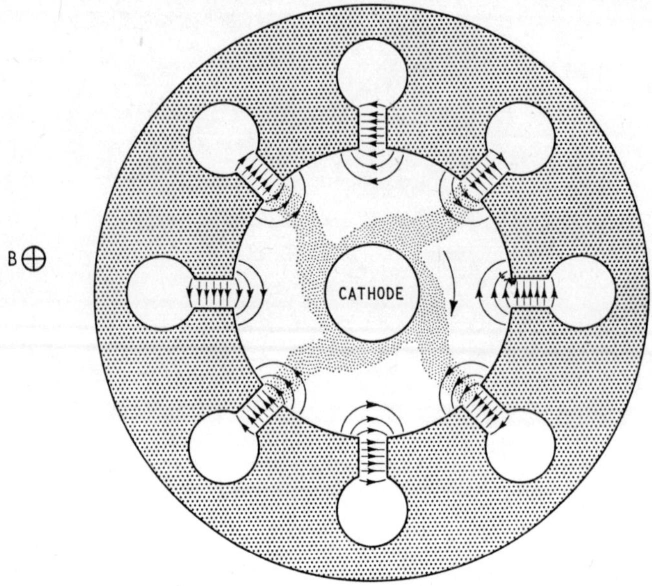

Fig. 12.12. *Space-charge and electric fields in an oscillating magnetron*

that they were first recognized. It has been shown (Equation 9.4) that the propagation of the *m*th space harmonic is given by

$$\beta_m = \beta_o \pm \frac{2\pi m}{p}$$

where p is the pitch of the circuit.

The number of wavelengths around the anode surface is $2\pi r_a/\lambda$, i.e. $\beta\, r_a$, hence if there are n' wavelengths of the *m*th space harmonic,

$$n' = \beta_m r_a$$

and

$$n = \beta_o r_a$$

hence

$$n' = n \pm m.\frac{2\pi r_a}{p}$$

But

$$\frac{2\pi r_a}{p} = N$$

the number of resonators around the circuit, hence

$$n' = n \pm mN$$

12.11.1 The (π–1) Reverse Mode

Usually only the first space harmonics ($m=1$) have amplitudes large enough to produce oscillations, and in the case of the π-mode, where $n=\frac{N}{2}$, putting $m=-1$ gives the same numerical value of n'. This means that the rotating space charge interacts equally well with either of the two rotating waves which form the standing wave. In the case of the $\pi-1$ mode, however, in an eight-segment block, for instance, $n=3$, but n' (for $m=-1$)$=5$, numerically. This means that one of the two rotating components of this mode can be excited by either a three-spoked space charge rotating in the same direction, or by a five-spoked space charge rotating in the opposite direction. The $n'=5$ possibility will, if the ($\pi-1$)-wavelength is close to the π-mode wavelength, have a lower operating voltage, and hence could be expected to give trouble.

12.12 Strapping

These difficulties make it very desirable to increase the mode separation by some means, and one way of doing this is by strapping, as illustrated in Fig. 12.7. This consists in joining together the tips of alternate vanes (which in the π-mode are at the same potential) by short loops or strips. These effectively add capacitance to the cavity in the π-mode, and hence reduce the π-mode frequency. This effect may be small, as with the wire strapping of Fig. 12.7(b), or relatively large, as with double-ring strapping (Fig. 12.7(a)), the blocks being said to be lightly or heavily strapped respectively. The effect of the straps on other modes is to make rather larger alterations in their resonant frequencies; their effect is equivalent to adding series inductance to the cavity, and the frequency is increased. By this means, therefore, the desired objective of wider mode separation is achieved, as indicated in Fig. 12.13. The introduction of strapping provided the first stable high-power pulsed magnetrons.

Movement of the straps also provides a means of tuning the anode block, i.e. of slightly altering the resonant frequency before the valve is completed, to obtain the desired operating frequency.

12.13 The Rising-Sun Anode System

An alternative approach is given by the so-called 'rising-sun' block design, shown in Fig. 12.7(e). Here the slow-wave structure

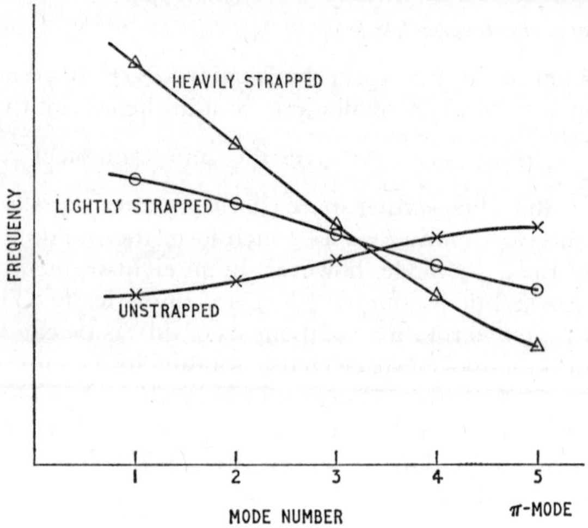

Fig. 12.13. *The effect of strapping on the mode-spectrum of a 10-segment magnetron*

has been made bi-periodic, consisting of alternate long and short cavities. This has the effect of splitting the mode spectrum into two regions, which are effectively the spectra of the small and large sets of cavities respectively, and the π-mode is where these two join, i.e. near the centre of the total mode spectrum. Fig. 12.14 shows how the relative mode frequencies of a uniform eighteen-slot block change as the relative depths of alternate slots is increased, and it can be seen that a spectrum of two sets of frequencies, both well separated from that of the π-mode is obtained.

Rising-sun construction is particularly convenient at the higher frequencies, where the sizes involved make machining increasingly difficult, and in the region above 10,000 Mc/s such blocks are commonly produced by hobbing. An example of a 35,000 Mc/s hobbed rising sun block is shown in Plate 12.2.

12.14 Types of Coupling

As indicated earlier, the relative loading on different modes can also affect mode selection, and various methods have been devised to produce loading on the otherwise unloaded members of the spectrum, particularly the $\pi - 1$ mode, in addition to the desired loading on the π-mode. Some forms of coupling are shown in Fig. 12.15,

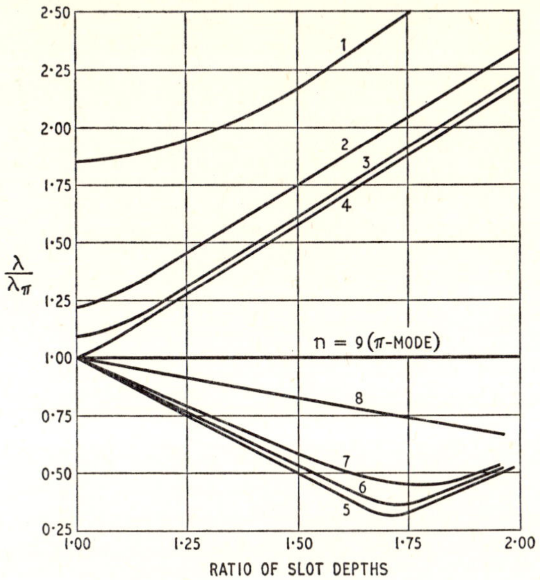

Fig. 12.14. *Effect of varying slot-depth ratio on the spectrum of an 18-segment block (after* OKRESS, 1961)

which show probe outputs fed from a loop in one cavity, or from connection to a strap, and loop outputs. One method of producing even loading around the block is that of Fig. 12.16, in which an axial output probe is fed by symmetrically arranged leads from the tips of alternate segments. This is a modern example of an otherwise unstrapped equal-segment block. The simplest output coupling is that shown in the drawing of Fig. 12.11, in which a slot leading from the back of one cavity is designed as a quarter-wave transformer to the external waveguide. The power then flows through a matched window, usually of glass or alumina, to the transmitter.

12.15 Back-Bombardment

One of the limitations to the maximum power of magnetrons arises, perhaps unexpectedly, from the cathode dissipation. When a magnetron is oscillating at a high power level, electrons emitted from the cathode in unfavourable phase are rapidly accelerated back to the cathode with relatively large energies. It would be expected that this is true of roughly half the total number of electrons emitted, and

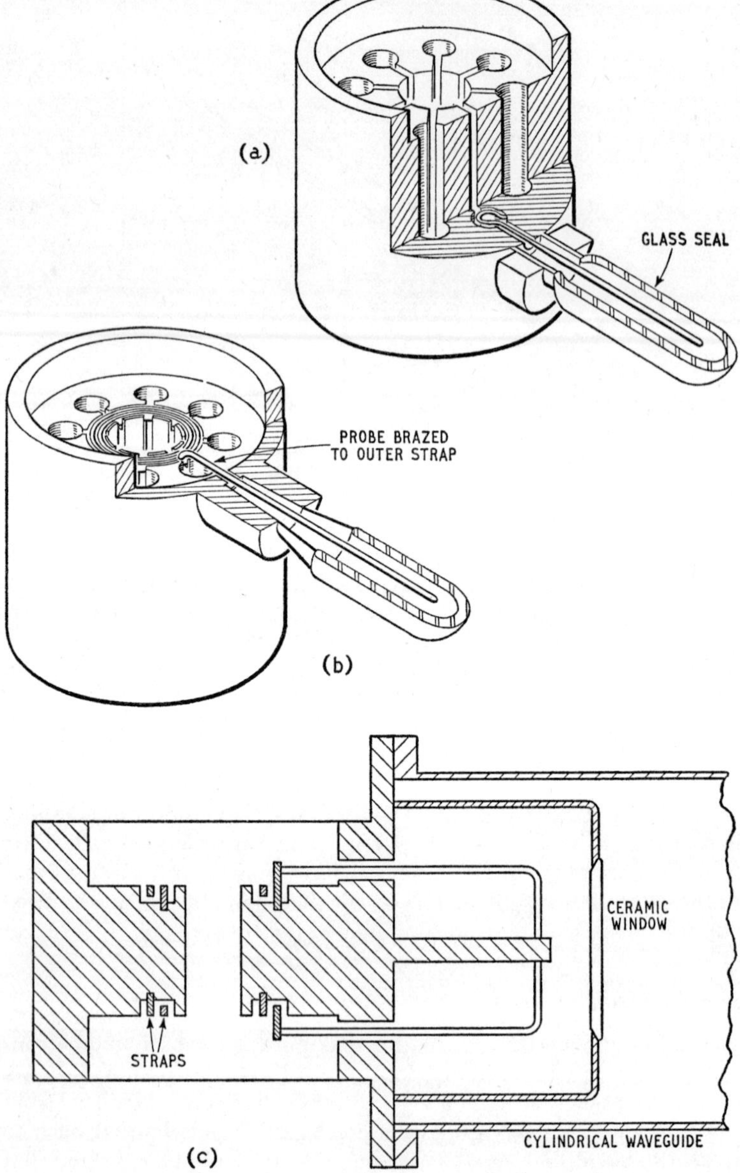

Fig. 12.15. *Some types of output coupling: (a) loop-fed probe output (b) strap-fed probe output (c) loop-fed from opposite straps*

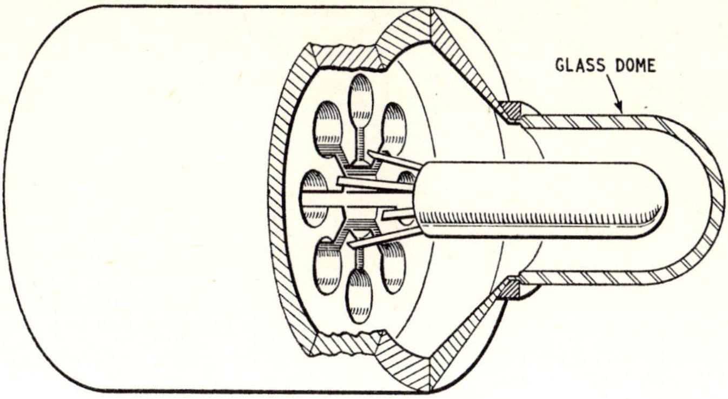

Fig. 12.16. *Waveguide probe fed symmetrically from alternate vanes*

in practice it is found that this back-bombardment can carry as much as about 8 per cent of the input power back to the cathode. Mean output powers may be of the order of several kilowatts, and hence the cathode heating may be increased by a few hundred watts. For this reason, high-power magnetrons are usually operated without any heater supply, and indeed will sometimes start from cold in this condition. The problems of cathode design in magnetrons, and to a lesser extent in Amplitrons are not, therefore, the usual ones of attempting to heat the cathode surface with the minimum amount of power, but of preventing it from overheating with large amounts of bombardment power.

12.16 Tuning

Various attempts have been made to provide externally operated mechanical tuning of magnetrons and many tunable magnetrons are now availale, although in general the maximum power output and efficiency are usually decreased as a result of the introduction of a tuning mechanism. The design difficulties of tuning arise from the fact that the resonant system is a multi-cavity one, and hence if any appreciable tuning range is to be achieved it must be done by something which affects all the cavities simultaneously. Some of the devices which have been used are shown in Fig. 12.17. Practical difficulties also arise from the fact that if the tuner is to be effective it must enter into either a high field or high current region. This readily leads to arcs being produced between the tuner and the body, which at the power levels involved are usually accompanied by the

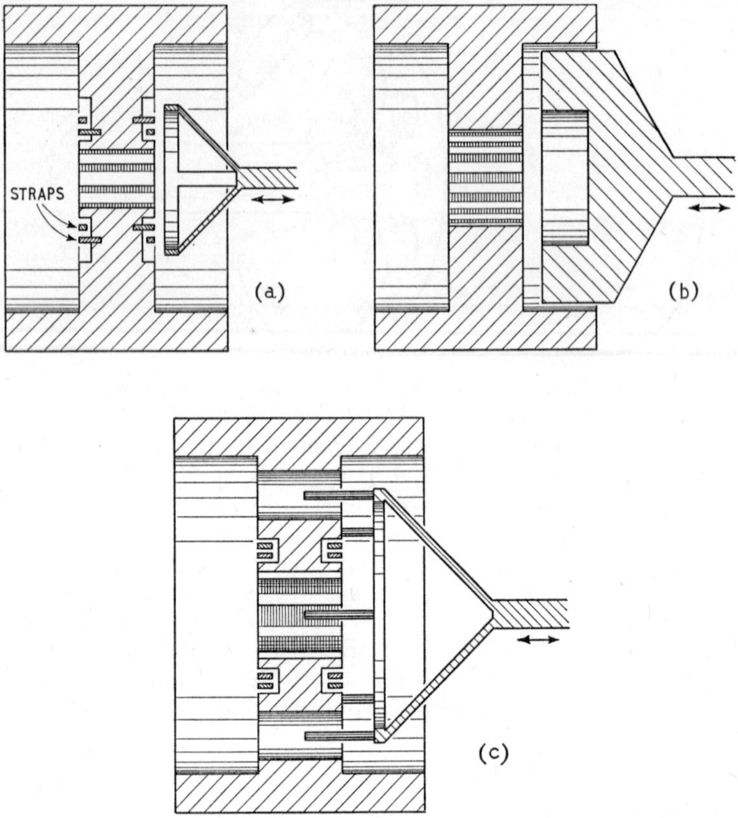

Fig. 12.17. *Magnetron tuning systems: (a) capacitive ring-tuner of strapped block (b) inductive ring-tuner of rising-sun block (c) 'crown-of-thorns' tuner in 8-segment hole and slot ring-strapped block*

rapid erosion of metal, thus permanently damaging the valve. There are also the difficulties of arranging moving mechanisms within the vacuum envelope, where lubrication is impossible, and this often leads to the life of the valve being limited by the operating life of the tuner.

12.17 Q-Factors

An equivalent circuit of a magnetron operating in one mode is that of Fig. 12.18. In this, L and C are the effective inductance and

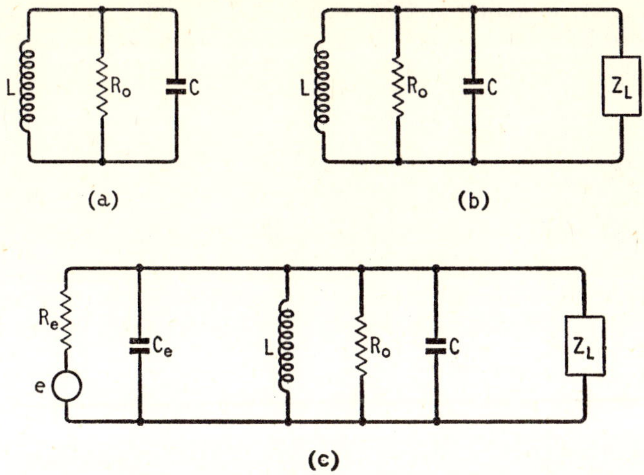

Fig. 12.18. *Equivalent circuits of magnetron for one mode: (a) cold unloaded magnetron (b) cold magnetron coupled to load (c) oscillating magnetron*

capacitance respectively, and R_o the effective shunt resistance due to the finite conductivity of the block, which is usually made of oxygen-free high-conductivity copper. Its resonant frequency is then

$$f_o = \frac{1}{2\pi\sqrt{LC}}$$

and its Q, defined as usual as

$$\frac{\text{Energy stored} \times 2\pi}{\text{Energy dissipated per cycle}}$$

can readily be obtained as

$$Q_o = \omega C R_o \qquad (12.11)$$

This can be measured on a block, for example, by using a very lightly coupled probe, with the normal output removed or short circuited. When the block is coupled through its output circuit to an external load the latter may be represented by a shunt impedance Z_L, which, if the load is a matched waveguide and the coupling circuit behaves, as is usually the case, as a lossless transformer, will be a resistance R_L. We then have a loaded Q, denoted by Q_L, which as above will be given by

$$Q_L = \omega C \left(\frac{R_o R_L}{R_o + R_L}\right) \qquad (12.12)$$

Another quantity sometimes used is the coupled or external Q, denoted by Q_E, which is

$$\frac{\text{Energy stored} \times 2\pi}{\text{Energy dissipated per cycle in load}}$$

and it is readily seen that

$$\frac{1}{Q_L} = \frac{1}{\omega C}\left(\frac{1}{R_o} + \frac{1}{R_L}\right) = \frac{1}{Q_o} + \frac{1}{Q_E} \qquad (12.13)$$

12.18 Effect of Loading

When the valve is oscillating it may be represented by Fig. 12.18(c), where R_e (which is negative) is the equivalent electron space-charge resistance, and C_e its effective shunt capacitance. This has the effect of reducing the resonant frequency below the cold frequency, and, of course, it is this lower frequency for which any frequency dependent circuits such as that of the output transformer must be designed. As with all oscillators, there is an optimum value of loading which produces most power in the external circuit. In practice, the coupling used is rather lighter than the optimum, for two reasons. Firstly, if the loading is very heavy, oscillation build-up time is increased, and hence mode-changing more likely to occur. Secondly, heavy loading increases the dependence of the operating frequency on the external circuit. In practice, the output waveguide cannot be maintained matched, particularly if, as is common, it is feeding a moving aerial, and so the load Z_L will often consist of a varying complex impedance. This will obviously affect the resonant frequency, and the effect is, in fact, sometimes made use of in tuning valves.

12.18.1 Pulling Figure

However, it is clear that the less tightly the valve is coupled to the output waveguide, the less this effect will be. It is defined in terms of a *pulling figure*, which is the maximum total variation in output frequency as a mismatch of constant voltage standing wave ratio of 1·5 is varied in phase. The choice of 1·5 is an arbitrary one, and the total effect of the load on the resonant frequency can be shown in a Rieke diagram, Fig. 12.19. In this diagram, which makes use of the Smith chart used in transmission-line analysis, the concentric circles represent lines of equal reflection coefficient or standing wave ratio, and the angle around the circle represents movement of the

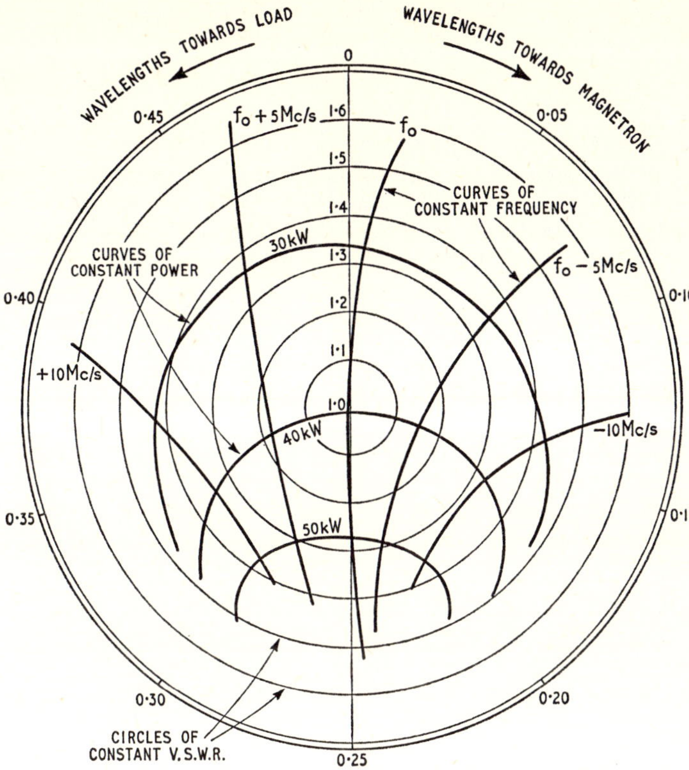

Fig. 12.19. *Performance shown on Rieke diagram*

mismatch towards or away from the valve, so that the effect of moving a mismatch obstacle along the waveguide is shown by going round the appropriate circle on the Rieke diagram. This then shows all the variations of frequency and power that a given mismatch can produce.

By considering the equivalent circuit diagram of Fig. 12.18, it may be shown that if Δf is the change in cold resonant frequency as the 1·5 mismatch phase is varied,

$$\Delta f = \frac{0{\cdot}417 f_0}{Q_E}$$

A typical figure of Δf at 3,000 Mc/s is about 5 Mc/s, and for this Q_E must be adjusted to be about 250. It must, of course, be less than

Q_o, which for a ring-strapped block at this frequency might be 1,000–2,000.

12.19 The Magnetron Performance Chart

The performance of a magnetron feeding into a matched load is usually shown by a performance chart or contour diagram, as shown in Fig. 12.20. This starts out by being a simple plot of operating voltage against current at various magnetic fields. On this

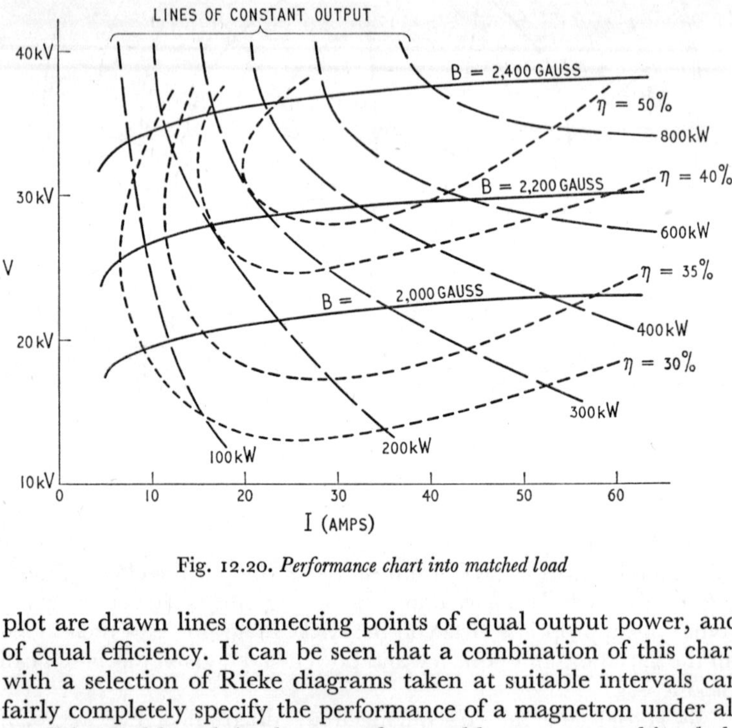

Fig. 12.20. *Performance chart into matched load*

plot are drawn lines connecting points of equal output power, and of equal efficiency. It can be seen that a combination of this chart with a selection of Rieke diagrams taken at suitable intervals can fairly completely specify the performance of a magnetron under all conditions, although in the case of a tunable magnetron this whole set would need to be repeated at various frequencies across its tuning range.

12.20 Magnetron Efficiency

It can be seen from the diagram of Fig. 12.20 that the efficiency increases as the magnetic field is raised and that ultimately the efficiency falls as the current is increased at any constant magnetic field. The electronic efficiency must always be given by

$$\eta = 1 - \frac{\frac{1}{2}mv^2}{eV_o}$$

and
$$v^2 = v_r^2 + v_\theta^2$$

At any given frequency, v_θ is constant $\left(=\frac{r\omega}{n}\right)$, hence the maximum possible electronic efficiency is

$$\eta = 1 - \frac{\frac{1}{2}\frac{mr_a^2\omega^2}{n^2}}{eV_o} \tag{12.14}$$

The only variable here is V_o, which is increased by using a high magnetic field, as indicated in Fig. 12.10 and this again indicates the value of using a high magnetic field. As the current is increased the radial component of velocity of electrons arriving at the anode also increases, and so the efficiency is decreased. Ultimately the current may increase so much that the r.f. fields can no longer contain the electron bunches, and the valve then drops out of oscillation. This occurs at the 'drop-out' current.

12.21 Uses of Magnetrons

By far the majority of magnetrons are high-power pulsed valves, most frequently used as the r.f. power source in pulsed radar systems. They have many advantages; they are the smallest valves available for any given frequency and power, they are relatively cheap and, since they determine their own operating voltage, they can be run from fairly simple power supplies. They cannot, however, readily be modulated in phase or amplitude within the pulse, nor can their frequency be quickly altered. They have much poorer signal-to-noise ratios than beam amplifiers, and in general their lives are shorter than those of beam-power amplifiers. As a result, they are now tending to be displaced from the military field where cost is less important than performance and the need is to increase the sophistication of performance, but will no doubt continue to be used for many years in civil systems, particularly in shipborne and airborne radar navigation aids, and for military applications where size and weight are all-important.

There are also c.w. magnetrons, which have particular applications in the field of radio warfare and jamming systems, although in this junction they have been displaced by high-power crossed-field backward-wave oscillators. There is, however, one use for c.w. magnetrons which is steadily increasing, and this is for microwave cooking, where their cheapness, small size and simplicity make them the best choice. These produce power of 1 to 3 kW at about 2,450 Mc/s, which is directed into the food to be cooked or reheated. Because the absorption of power takes place throughout the entire volume of the food, its temperature rises uniformly throughout, and joints of meat, for example, may be heated or cooked in minutes, which by the normal process of heating from the outside would take hours.

12.22 Injected-Beam Magnetrons

During the last few years there has been a growth of injected-beam magnetrons, in which the rotating beam is generated separately in an electron gun, and then passed into the magnetron interaction space. By this means the space-charge angular velocity is externally generated and controlled. This helps to maintain stable operation under very heavily loaded conditions and is of use when it is desired to vary the frequency by external tuning or by altering the space-charge conditions. The former behaviour can be immediately understood, since the external Q_E of such a valve will be very low and hence its pulling figure very high, e.g. about one-half f_0. The electronic tuning behaviour is rather more complex, and depends on the variation of the electron space-charge impedance as conditions are varied.

12.23 Voltage Tuning

It has already been noted that the operating frequency of a magnetron is slightly different from its cold resonant frequency, due to the reactive part of the space-charge impedance, as indicated in Fig. 12.18. The effect of this impedance varies with power level, and in high-power magnetrons relatively lightly coupled to their loads causes *frequency pushing*, which is the variation of frequency with operating current. In such valves the voltage remains fairly constant as the input power is varied, and the extra power consumption occurs predominantly as an increase of current. The principal mean direction of electron flow is always azimuthal.

If the current is limited, however, as is the case if it is supplied by an external gun, then the voltage will rise instead, and the frequency pushing effect is replaced by a voltage-tuning effect. As in the case of externally tuned magnetrons, the valves are very heavily loaded by their external circuit, which makes tuning over a large range possible and frequency ranges of as much as 2 : 1 are obtained.

Magnetrons like this, which are much less efficient than the high-power valves, are of use as low-power generators, as a higher power

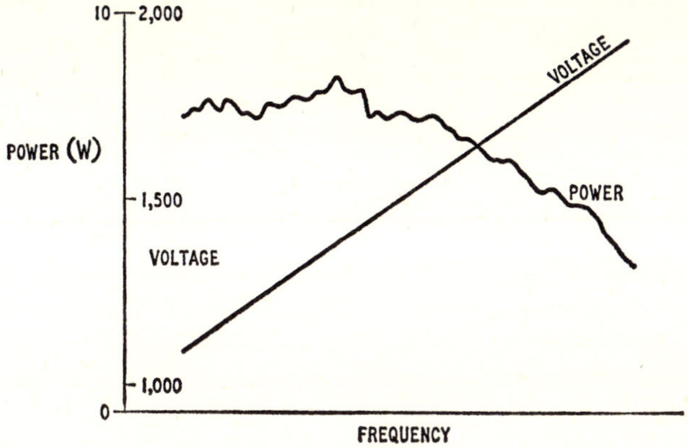

Fig. 12.21. *Form of voltage and power variation with frequency of voltage-tuned magnetron*

alternative to O-type backward-wave oscillators. Their chief characteristics of frequency against voltage, and power output against frequency are shown in Fig. 12.21. Their construction is quite different from that of high-power magnetrons. The anode system is made in an interdigital form, as shown in Fig. 12.22, and the entire

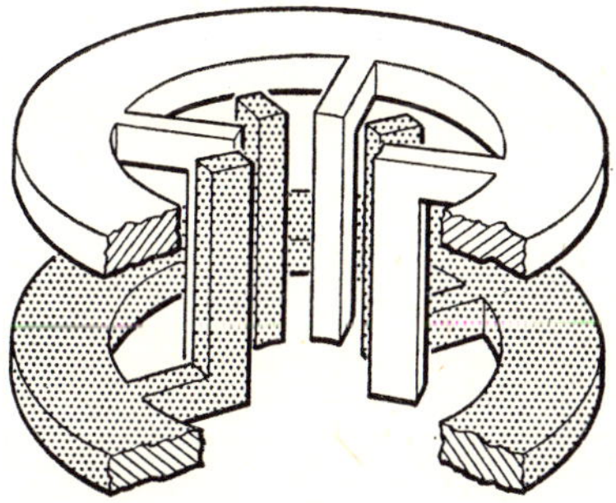

Fig. 12.22. *Interdigital anode for voltage-tuned magnetron*

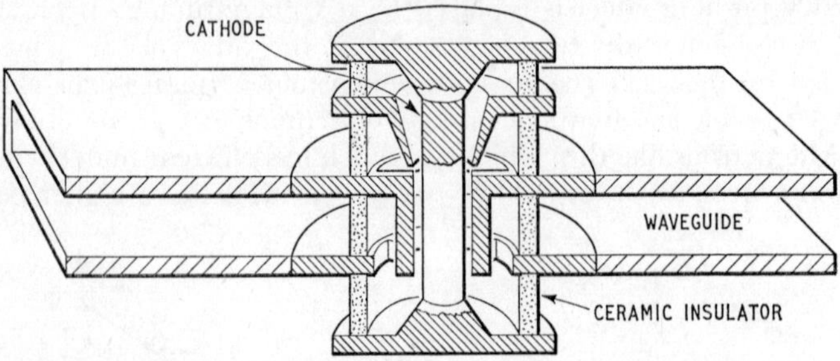

Fig. 12.23. *Mounting of voltage-tuned magnetron*

valve is made so that it can be placed across a waveguide or transmission line, thus achieving the heavy loading, as shown in Fig. 12.23.

BIBLIOGRAPHY

OKRESS, E., *Crossed-Field Microwave Devices, Vols. I and II,* Academic Press (1961).
COLLINS, G. B., *Microwave Magnetrons,* M.I.T. Radiation Laboratories Series No. 10 (1948).

Chapter 13

NOISE

13.1 Thermal Noise

A simple description of noise is that it is what we do not wish to hear or, more generally, receive. In this it is opposed to signal, which is what it is desired to receive. It is, of course, immediately apparent that one man's noise is another man's signal, and that much of the noise received in many channels of communication, such as telephone, radio, television or merely shouting to one another in a crowd, is man-made, arising from electric machinery, aircraft, road traffic, railways, transitor radios, etc. There is also present in the atmosphere a large amount of electrical disturbance due to static discharges such as lightning, which gives rise to noise in receivers, particularly at radio-communication frequencies. However, even when all man-made and atmospheric sources of interference are removed, the performance of equipment is still limited by noise generated in its components; this chapter is about this latter noise and ways in which it can be reduced.

13.1.1 Thermal Noise in Aerials and Feeders

If we suppose that we have a completely noise-free ideal receiver connected by a lossless transmission line to a perfectly matched aerial, we find that it is not possible to increase the receiver sensitivity indefinitely by increasing its gain, and that ultimately the receiver becomes saturated with noise being received by the aerial. This is due to the fact that the molecules and electrons of all matter are in general not stationary, but are continuously vibrating in thermal motion, and are consequently emitting radiations. If the temperature of an object is very low, there is less energy in these vibrations, and their effective frequency range is reduced, but as the temperature is increased the vibrational energy also increases (indeed, this statement might equally well be written the other way round) and hence the energy being radiated increases and the centre of its frequency spectrum is shifted upwards. Since the radiation is due to each particle vibrating more or less independently, the radiation is not *coherent*, i.e. not of repetitive waveform like an oscillation output, but random and is referred to as noise, from analogy with the audible effect at low frequencies. Because of its origin, this type of noise is known as *thermal noise*. It covers the entire frequency range, with an approximately constant level over the radio-frequency region. This constant level is referred to as *white noise* in the same way as an approximately constant level of light over the visible region is called white light. If the temperature is raised to about 900°K, radiation at frequencies as high as those of red light (about 4×10^8 Mc/s) starts to appear in appreciable quantity, and we say that the object is red-hot. If it is made still hotter, we can see the widening of the emitted frequency spectrum as a change in colour, although only a very small part of the total energy radiated is in the visible spectrum.

The power radiated in a small range δf at a frequency f is given by

$$P(f) = \frac{hf\delta f}{[e^{hf/kT} - 1]} \qquad (13.1)$$

which is one form of Planck's radiation formula, a well-known result in thermodynamics, where h is Planck's constant $= 6{\cdot}63 \times 10^{-34}$ Js, k is Boltzmann's constant $= 1{\cdot}38 \times 10^{-23}$ J/deg. and T is the absolute temperature.

It is apparent that when we are concerned with frequencies in the microwave and radio regions, below, say, 100,000 Mc/s, and temperatures above freezing point, i.e. 273°K, then hf/kT will be small, (less than 0·0175). By expansion of the exponential part of $(e^{hf/kT} - 1)$ we then see that this expression effectively becomes hf/kT, and Equation 13.1 reduces to

$$P(f) = kT\delta f \qquad (13.2)$$

which is independent of the frequency. T is now the mean absolute temperature seen by the aerial. The temperature of things at the surface of the earth is usually taken as 290°K, and this conveniently gives $kT = 4 \times 10^{-21}$ J. We can thus express Equation 13.2 in a different way, and say that the noise energy arriving at the aerial is 204 dB below 1 watt per cycle of bandwidth. This imposes a limit on receiver sensitivity, since the information in any signal arriving below this level of energy is difficult to detect.

13.1.2 Thermal Noise from Resistors

In the same way, it can be shown that the thermal movement of molecules and electrons in a resistor also gives rise to a maximum electrical noise power of kTB, where we use B as the bandwidth in c/s, instead of δf. This maximum noise power is available from the resistor when it is connected to a load equal to its own resistance. This means that in a circuit consisting only of two equal resistors connected in parallel there is a constant flow of noise currents through each of them due to the noise power available from the other, which can be considered to be due to noise e.m.f.'s from each of them. From this it follows that any circuit component causing loss or dissipation, such as waveguide or coaxial feeders, coils, resonators, joints, in fact everything which is not ideal and completely loss free, will give rise to thermal noise.

This thermal noise in electrical circuits is random in its waveform and has no d.c. component; the voltages and currents are therefore expressed as mean square values, denoted by $\overline{V^2}$ and $\overline{i^2}$ respectively. Thus, as in Fig. 13.1(a) a resistance of value R ohms can be considered to have a noise voltage generator of value $\sqrt{\overline{V^2}}$ in series with it. The current fed into a second resistance R is then $\sqrt{\overline{V^2}}/2R$ and the power developed in the second resistance is $\overline{V^2}/4R$. But we saw above that this is kTB, so that the mean square voltage due to any resistance R is

$$\overline{V^2} = 4kTRB \text{ (volts)}^2 \tag{13.3}$$

For example, a resistance of 1,000 Ω at room temperature will produce 4 μV r.m.s. in a frequency band of 1 Mc/s. This mean square voltage may be considered to be a circuit element and be manipulated by ordinary circuit theory. In particular, by Thevenin's theorem, an alternative representation of a resistor is as Fig. 13.1(b), with a constant current generator of value

$$\overline{i^2} = 4kTGB \text{ (amps)}^2 \tag{13.4}$$

in parallel with the conductance $G(=1/R)$.

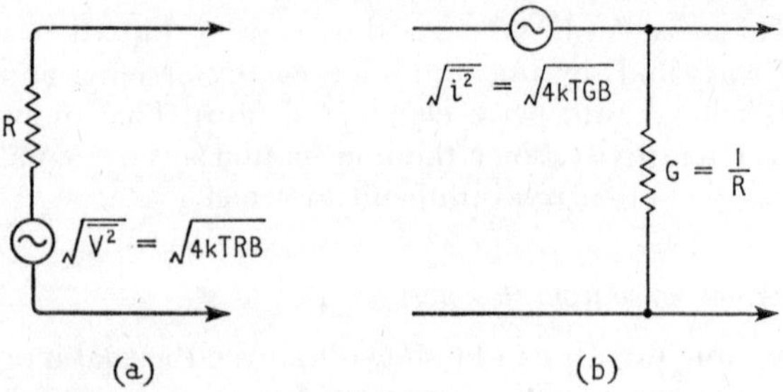

Fig. 13.1. *Alternative equivalent noise circuits for a resistor*

13.2 Noise Figure of Amplifiers

Unfortunately, it is only rarely that thermal noise becomes a limitation on performance. The limit more often arises in the much larger quantities of noise generated in crystal mixers or valve or solid-state amplifiers. The worsening of the noise is usually defined by the ratio

$$F = \frac{\text{Signal/noise ratio at input}}{\text{Signal/noise ratio at output}}$$

which is called the *noise figure* or *noise factor* and is usually given in decibels.

13.3 Overall Noise Figure of Amplifiers in Cascade

It is often convenient to represent a real noisy amplifier as if it consisted of an ideal noiseless amplifier with a hypothetical noise source connected to its input as shown in Fig. 13.2. Then if there is

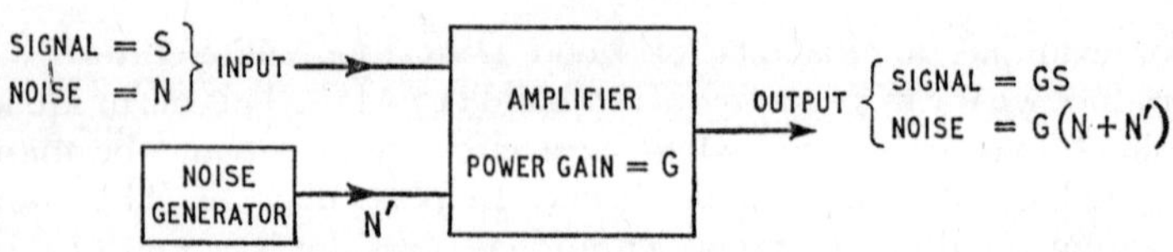

Fig. 13.2. *Ideal representation of amplifier with noise figure* $F = \left(1 + \frac{N'}{kTB}\right)$

an input signal at some level S watts, and the noise level accompanying this signal is N watts, the input signal to noise ratio is S/N.

Also feeding into the amplifier there is the hypothetical noise source of N_1 watts. With an amplifier power gain G the output signal level is now GS watts, while the output noise level is $G(N+N_1)$ watts. The signal to noise ratio at the output is therefore $S/(N + N_1)$, and has been degraded by the ratio

$$\frac{N + N_1}{N} = 1 + \frac{N_1}{N}$$

which is therefore the noise figure. The incoming noise N, if the amplifier is a receiving input stage, will usually be the basic thermal noise, kTB watts, where for terrestrial communication T is usually taken at 290°K. Under these circumstances, if we take kTB as a unit of noise, then $N=1$ and N_1 can be taken as a multiple of this unit, and the noise figure becomes $(1+N_1)$.

Similarly, two amplifiers in cascade, with gains G_1 and G_2, can be considered as consisting of two noise-free amplifiers, with respective hypothetical noise generators of N_1kTB and N_2kTB feeding into their inputs as in Fig. 13.3. We see that the signal to noise ratio is then degraded overall by the ratio

$$1 + N_1 + \frac{N_2}{G_1}$$

or expressing this in terms of the noise figures F_1 and F_2 of the two amplifiers, the overall noise figure

$$F = F_1 + \frac{F_2 - 1}{G_1}$$

It is thus evident that provided the gain G_1 of the input amplifier is large enough the noise contribution from the second amplifier is relatively negligible. To obtain maximum sensitivity, therefore, high-gain low-noise input stages are required.

13.4 Excess Noise Temperature

For the above amplifier the noise power which has been added to the basic thermal noise is

$$\left(N_1 + \frac{N_2}{G_1}\right) kTB$$

Sometimes the amplifier performance is quoted in terms of this, which is referred to as the *excess noise*. For low-noise amplifiers this

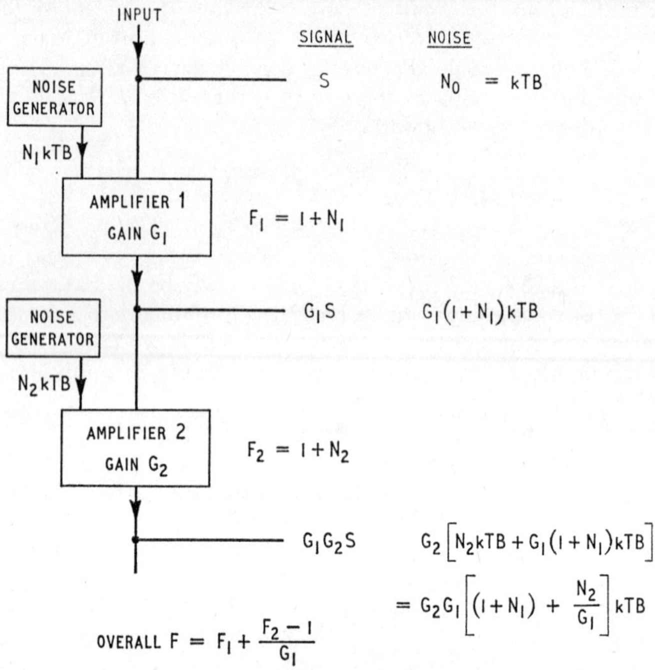

Fig. 13.3. *Noise performance of two amplifiers in cascade*

step is carried farther, and the performance is given in terms of *excess noise temperature*. In the above, if $N_1 T = T_1$ and $N_2 T = T_2$, the excess noise temperature is

$$T_1 + \frac{T_2}{G_1}$$

With a source at a temperature T_s rather than 290°K, the overall noise temperature of the receiver may then be said to be

$$T_s + T_1 + \frac{T_2}{G}$$

In all the valves described in earlier chapters there is a conversion of d.c. energy from a power supply into r.f. energy, and this is carried out by an electron beam or a flow of electron current from a cathode. This cathode is the major source of irreducible noise in these valves.

13.5 Noise from Cathodes; Shot Noise and Rack Noise

When describing valve operation we often treat the cathode current as if it were a tenuous fluid, capable of being reduced to any smooth level, and without any lumps in it, and having only a smoothly varying velocity across any section. In reality, the current consists of individual electrons (of charge e coulombs) which, by the very process of thermal emission have a considerable spread of velocities, actually lying in the range zero to infinity, but mostly with a low value. For instance, for a typical cathode temperature of 1,000°K the mean velocity corresponds to 0·068 V. Also the electrons are not emitted continuously, but with a random variation in time which is quite independent of the randomness of velocity.

The noise quantities which result on the beam can be calculated, and it can be shown that over a bandwidth B the mean square value of the current fluctuations arising from the random discontinuous emission of the current (I amp) leaving a cathode is

$$\overline{i^2} = 2eIB \text{ (amp)}^2 \qquad (13{\cdot}5)$$

Thus a cathode current of 1 mA will have a r.m.s. fluctuation component of value 0·018 μA in a frequency band of 1 Mc/s. Similarly the mean square velocity fluctuation of the emitted current due to random emission velocity is

$$\overline{u^2} = \left(1 - \frac{\pi}{4}\right) \frac{\eta}{I} 4kT_c B \text{ (metres/sec)}^2 \qquad (13{\cdot}6)$$

where T_c is the effective cathode temperature, and η ($=1{\cdot}76 \times 10^{11}$ C/kg) is the ratio of the charge of the electron to its mass. The noise arising from the current fluctuation is known as *shot noise*, while that from the velocity distribution is often called *Rack noise*, after its original calculator.

13.6 Noise in Space-Charge Control Valves

In diodes, triodes, etc., it is the current fluctuation which is of major importance, and it is usually observed as a voltage fluctuation across an impedance in series with the valve. For example, from Equation 13.5, the mean square voltage fluctuation across a resistance R Ω in series with a temperature-limited diode passing a current I amp is given by

$$\overline{V^2} = 2eIR^2B \text{ (volt)}^2 \qquad (13{\cdot}7)$$

Thus a diode current of 3 mA passing through a resistance of 1,000 Ω

will produce 100 μV of noise over a bandwidth of 10 Mc/s. This compares with the 13 μV of thermal noise from the same resistance.

This voltage fluctuation will be constant for all frequencies where the transit angle $\omega\tau$ (Chapter 5) is small. When, however, $\omega\tau$ becomes greater than about $0{\cdot}2\pi$ rad, then, as in the case of the diode impedance, the noise per unit bandwidth becomes smaller. Since the total diode current is the random sum of a large number of pulses like that shown in Fig. 5.6, the reduction in noise amplitude may be said to occur at frequencies where a Fourier analysis of the pulse waveform gives lower values to the amplitude of the components.

A more important case is that of space-charge-limited cathode current where it is found that even at low frequencies the fluctuations in a given anode current are considerably less than for the temperature-limited case. This arises from the fact that as each electron passes through the space charge it causes a local depression of potential at the potential minimum which prevents other electrons from reaching the anode. The overall effect is to reduce the average effective charge per electron reaching the anode to a value Γe. Since e appears in Equations 13.5 and 13.7, both explicitly, and in the current ($I=ne$ where n is the average number of electrons leaving the cathode per second), the overall effect is large, the shot-noise current becoming

$$\overline{i^2} = \Gamma^2 2IeB \tag{13.8}$$

Practical values for Γ^2 lie between 0·05 and 0·2 for cathodes used in receiving type triodes and pentodes. Again, at higher frequencies, the noise output becomes smaller. For space-charge-limited diodes, as also for high-frequency temperature-limited diodes, the calculation of the noise voltage developed across a series impedance must take into account the shunting effect of the diode impedance.

In low-frequency triodes, when used in their normal manner, with the grid biased negatively so that no grid current flows, the situation is similar to that in a diode, and a shot-noise current, as given by Equation 13.8, is fed into the anode load. Since the anode-cathode impedance is higher than for the diode, its shunting effect is correspondingly smaller.

13.6.1 Partition Noise

In low-frequency pentodes additional noise arises from the sharing of the current between the screen grid and the anode. Even if the electrons were emitted uniformly in time from the cathode, there would still be random collection by the screen grid due to the random point of emission. The resulting *partition noise* current increases the overall fluctuation $\overline{i_a^2}$ in the anode current I_a from the value given by Equation 13.8 to

$$\overline{i_a^2} = \left(\frac{I_s + \Gamma_k^2 I_a}{I_k}\right) 2I_a eB \tag{13.9}$$

where Γ_k^2 refers to the cathode current I_k and I_s is the screen current.

The shot-noise current in the anode circuit of a pentode will be reduced at high frequencies by the finite transit angle, although in practical valves the effect will not be large over the working range, since the transit angle must be made small to enable the valve to work at all. The partition-noise current will, to the first order, not be affected.

13.6.2 Induced Noise

A new effect arises, however, in the control-grid circuit from the fluctuations in the induced-grid current. Roughly the induced-noise current in the grid circuit is that which would be produced as thermal noise by the transit-time input conductance (Chapter 5), and thus increases as the square of the operating frequency. This noise current will develop a noise voltage across the grid-circuit impedance and therefore produce noise current at the anode. Although this noise current and the shot-noise current arise from the same electrons, experiment shows that only a partial correlation exists between the two, and only a little reduction in total noise current can be obtained by adjustment of the magnitude and phase angle of the grid circuit impedance.

Triodes and pentodes thus cannot be designed for low noise at a particular frequency, but always give a noise performance steadily deteriorating with increase of frequency. In low-noise receivers using such valves, the first stage will always use a triode, and, with the best of these, noise factors of 3 dB at 100 Mc/s rising to 9 dB at 1,000 Mc/s can be obtained. Having obtained some initial amplification, slightly worse noise performance can be tolerated for the following stages, and for frequencies up to a few hundred megacycles per second these stages will use r.f. pentodes in order to obtain higher gain. At higher frequencies, up to, say, 1,000 Mc/s, triodes will have to be used throughout. At this frequency and above beam-type amplifiers become more appropriate.

13.7 Low-Noise Travelling Wave Tubes

In *O-type amplifiers*, when discussing the effects of the noise fluctuations at the cathode (or the potential minimum just in front of it), it is convenient to use again the concept of space-charge waves. The two types of fluctuations described, i.e. shot noise and velocity fluctuation, each excite space-charge noise waves which propagate along the beam from the cathode. Being noise waves, they do not have a repetitive waveform. Also, because the times of emission of the electrons are quite independent of their velocities there is no correlation between the two sets of space-charge waves, i.e. they cannot on the average cancel or augment each other. If nothing is done about them, these waves will in turn excite circuit waves when the beam enters the slow-wave circuit—or cavity responses in the case of klystrons—and thus give rise to amplified noise outputs over the bandwidth of the valve in question. In fact, if simple convergent flow guns are used in these tubes, the initial noise becomes considerably amplified, and noise figures of 25–30 dB result. Consequently most microwave valves that are not specially designed as low-noise amplifiers have noise figures of at least this value.

The most widely met electron-beam low-noise amplifier in which the beam energy is used to provide that amplification is the *low-noise travelling wave tube*, which in the principal aspects of its construction is similar to higher-power travelling wave tubes, but in which particular attention is paid to reducing the noise figure. It has already been noted that the place where low-noise performance is critical is most usually at the input to a receiver, so that low-noise t.w.t.'s are usually very low-power devices.

The problem of reducing excess noise to the minimum in these valves obviously starts at the cathode, where the noise originates, and special techniques are used. Firstly the cathode surface must be as smooth and well defined as possible. This has been shown experimentally to be important, and reasons for it may readily be conjectured. At the edge of a badly defined emitting area, for instance, temperature-limited emission might be expected to occur, which gives rise to greater noise than occurs in the normal space-charge-limited condition. Next, the cathode must be capable of giving a high level of emission at the lowest possible temperature, since the temperature of the cathode is reflected in the effective noise temperature of the beam. For these reasons low-noise cathodes have much more attention paid to their preparation and processing than is common in most higher-power devices. The cathodes are nearly always oxide coated, but the coating is often laid down by more refined methods than the spraying in suspension which is commonly met with. Methods such as printing, centrifuging or cataphoresis are used, the resultant coating having a greater density, a smaller thickness and a lower surface irregularity than in power valves. The cathodes may run at temperatures in the region of 650°C, rather than the normal range of 750–800°C.

Next, it has been found that the best form of cathode is not always that of a disc, as might be expected in a non-convergent immersed-flow gun, but that a better noise performance is often obtained if the cathode is annular and the coating extends over its edge and a little way down the side, so that to some extent it behaves as a magnetron cathode. Even where disc cathodes are used, the strongly divergent field produced by the first gun electrode is such that most of the emission comes from the edge and the beam is annular in form close to the cathode. The sizes involved are often so small that an annular form of the cathode itself is impractical. The cathode is not designed to give uniform emission density from the whole of its surface, and the electrode shapes in the gun are arrived at entirely by the criterion of noise performance.

13.7.1 Drift Region near the Cathode

The gun is placed in an intense axial magnetic field, many times higher than the Brillouin field appropriate to the final velocity, whose function seems to be to effectively constrain the beam to be only one-dimensional and to maintain a high space-charge density. It is then desirable for the beam to drift for some time (although as it is only moving relatively slowly it does not go very far) at an overall velocity of the same order as the thermal velocities which are present. It has been shown experimentally and by computer and theoretical studies that in this condition the space-charge wave propagation on the beam is quite different from that on a normal beam, and that the noise of the beam may actually decrease.

In this region, although the total current is very small (usually less than 1 mA), the low velocity causes the effective plasma frequency to become very high, and one way of looking at this part of the noise-reduction process is that if the plasma frequency becomes comparable to or higher than the signal frequency the concept of the beam as a lossless transmission line breaks down and the noise waves are not able to propagate. In addition, because the overall motion is at a velocity of the same order as the thermal-noise velocities which are present, the beam must be regarded as a multi-velocity beam, composed of many beams with a relatively wide spread in their velocities. This would give multi-phase propagation, and signals initially excited would soon become undetectable. Both these approaches, although giving some clue to the noise-smoothing behaviour, suffer because they use concepts not really appropriate to an electron cloud of this sort, and it might be expected that a statistical dynamic approach might be more fruitful.

All this part of the process takes place in the region immediately adjacent to the cathode, typically within about 0·1 mm from the cathode surface, and effectively results in a low-noise, slow-moving electron cloud, from which the beam for use in the tube is drawn.

The requirement for this condition also imposes a limit on the power level of the tube, since it is very difficult to draw currents of more than a few hundred microamperes from the cathode without using voltages which would prevent the existence of the low-velocity drift region. The amount of current drawn is controlled by both the first and second electrodes of the gun, which are both held at low voltages as indicated in Fig. 13.4.

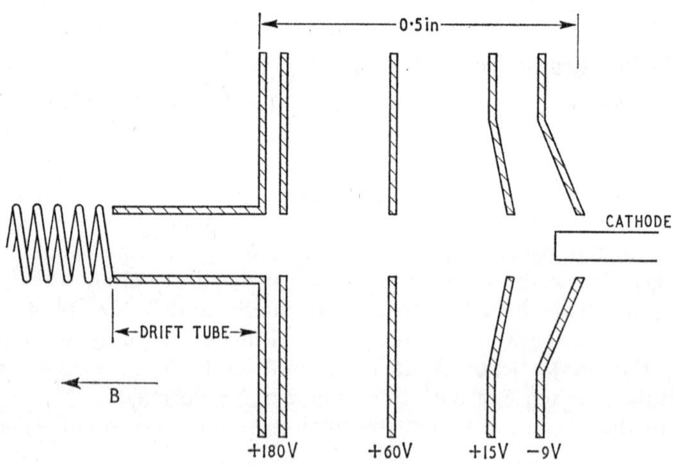

Fig. 13.4. *Low-noise electron gun*

13.7.2 Beam Impedance Transformation in the Electron Gun

In considering the behaviour of the low-velocity beam leaving the cathode region, we return to the idea of its consideration as a lossless transmission line, as introduced in Chapters 6 and 7. The relation between the current modulation $J_{\sim}$ and the velocity modulation $u_{\sim}$ was there shown to be (Equation 6.30).

$$\left|J_{\sim}\right| = \frac{\omega}{\omega_p}\frac{J_o}{u_o}\left|u_{\sim}\right|$$

Corresponding to the velocity variation $u_{\sim}$, there will be an effective voltage variation $V_{\sim}$, called the *kinetic voltage*, which is defined by

$$(u_o + u_{\sim})^2 = 2\eta(V_o + V_{\sim}) \qquad (13.10)$$

or, since $u_{\sim}$ and $V_{\sim}$ are small quantities, and $u_o{}^2 = 2\eta V_o$,

$$V_{\sim} = \frac{u_o}{\eta}u_{\sim} \qquad (13.11)$$

Hence

$$\frac{u_{\sim}}{u_o} = \frac{V_{\sim}}{2V_o} \qquad (13.12)$$

Using this, Equation 6.30 becomes

$$\left|J_{\sim}\right| = \frac{\omega}{\omega_p}\frac{J_o}{2V_o}\left|V_{\sim}\right| \qquad (13.13)$$

This relation between current and voltage variation defines an effective characteristic impedance Z_o for the beam, given by

$$Z_o = \frac{\omega_p}{\omega}\frac{2V_o}{J_o} \qquad (13.14)$$

The effective power of the wave is then, as in a real transmission line,

$$W = \tfrac{1}{2}J_{\sim}\,V_{\sim} \qquad (13.15)$$

and this is called the *kinetic power*. It will be apparent from Equation 6.16 that in a slow wave the kinetic power is negative, while in a fast wave it is positive. It is effectively the power added to the d.c. beam by the presence of a r.f. wave. When this r.f. input is due to noise, the added power is called the *kinetic noise power*. For an electron beam with uncorrelated density and velocity input fluctuations given by Equations 13.5 and 13.6, it can be shown that the effective total value of this quantity, now denoted by the symbol S, is given by

$$S = \left(1 - \frac{\pi}{4}\right)^{\frac{1}{2}}\frac{kT_c}{\pi} \qquad (13.16)$$

It must be remembered that here T_c is not now the cathode temperature, but some effective temperature produced by the noise-smoothing mechanism close to the cathode, described above.

If fast and slow waves are present together, we have standing waves on the beam, and an examination of the boundary conditions at the start of the t.w.t. helix shows that, for minimum noise performance, the level of the standing wave ratio and the distance of the current minimum from the start of the helix can both be chosen to give the best result. This is effected by (*a*) controlling the axial variation of Z_o, which must, of course, be changing through the gun, since ω_p and V_o are both varying, and (*b*) arranging a suitable transit angle from the last current minimum, which is most conveniently done by having a drift space between the gun and the helix. The correct variation of Z_o through the gun is effected by having several electrodes in the gun to give the appropriate d.c. potentials, instead of only one as is the case in power valves, and the resulting gun is consequently a more complicated structure. An example of the geometry which results is shown in Fig. 13.4.

13.8 T.W.T. Minimum Noise Figure

If the gun design is adequately carried through, and the gun does not itself introduce any further noise into the beam, either by electron lens effects or partition noise through imperfect focusing, then the minimum noise figure of the resulting travelling wave tube will be

$$F_{min} = 1 + \frac{2\pi}{kT_o}(S - \Pi)\left(1 - \frac{1}{G}\right)\left(1 + \frac{d}{x_1}\right) \qquad (13.17)$$

where G is the gain of the tube, x_1 is the constant of increase of the growing wave (see Chapter 7) and d is the corresponding constant of decay in the cold tube. Π is a constant that provides for possible correlation between the velocity and density fluctuations at the effective cathode. If, as is often assumed, no such correlation exists, and the gain G is made large, and the helix loss (hence d) is small, then the minimum noise figure becomes

$$F_{min} = 1 + \frac{2\pi S}{kT_o}$$

which with Equation 13.6 gives

$$F_{min} = 1 + \sqrt{(4 - \pi)}\,\frac{T_c}{T_o} \qquad (13.18)$$

Taking T_c as 1,000°K (cathode temperature), and T_o as the usual

290°K, this gives a minimum noise figure of 6·2 dB. In fact, considerably lower noise figures are obtained in practice, the lowest to date being about 1·0 dB, for a tube in which the conductors were at the temperature of liquid helium in order to minimise *d*. This gives an idea of the effectiveness of the noise-smoothing properties of the low-velocity drift region, since the corresponding beam temperature is only 81°K.

13.9 Other Sources of Noise in T.W.T.'s

After reducing the effect of the beam thermal noise to a minimum, it remains to ensure that no noise is added. Other possible noise sources, in approximately decreasing order of importance, are:

1. Reflected electrons from the collector, which can carry noise back from the output to the input.
2. The presence of excessive quantities of ions in the beam. These are excited by the beam into noise-like oscillations at frequencies up to a few megacycles/second. These oscillations modulate the beam and give rise to low-frequency noise modulation on a r.f. signal.
3. Beam interception at the input end of the tube, which will introduce partition noise into the beam, and may also produce secondary electron emission, which, for a given current, usually gives rise to more noise than the primary emission.
4. Loss in the input r.f. couplers and in the early part of the helix before the growing wave has become large enough to make the thermal noise from these sources negligible. Thermal noise also arises from the attenuator, so that, in a low-noise tube, it is often situated farther away from the input than is normal in a power tube.

Reflected electrons are avoided by arranging for the beam to diverge as it enters the collector, so that the surface is struck obliquely and by coating the collector surface with a material with a low secondary emission coefficient, such as carbon. The effect is also reduced by running the collector at a higher voltage than the helix.

Ion-modulation noise, which can also have a disastrous effect in high-power tubes, is avoided by careful attention to the selection and processing of materials, the obtaining of very good vacua, and by the provision of an ion drain in the tube, e.g. by again running the collector at a higher voltage than the helix.

These two effects can cause enormous increases in the noise output from travelling wave tubes, and give rise to signal-to-noise ratios corresponding to noise figures of as much as 70 dB.

The methods of focusing the beam in low-noise tubes are similar to those for other t.w.t.'s, but may be complicated in the case of the lowest-noise-figure tubes by the requirement for a very large magnetic field in the electron gun, and there has to be, as is so often the case, a compromise between best performance and greatest convenience. The best noise figures, of less than 3 dB, are obtained over bandwidths of the order of 30 per cent from tubes operating in large-field solenoids. Over greater bandwidths the noise level rises towards the edges of the band, and octave-bandwidth tubes with noise figures of less than 5 dB below 2,000 Mc/s, or less than 8 dB at 9,000 Mc/s, can be obtained. If periodic permanent magnet focusing (Chapter 10) is used, there is usually some deterioration of noise figure, and a compromise is often obtained by the use of single field-reversal permanent-magnet focusing.

13.10 Dynamic Range

A requirement often accompanied by that for low noise is for a mimimum saturated power. The reason for this can be seen by considering a practical case. If we have a low-noise tube with a noise figure of 8 dB, and a gain of 40 dB, then the noise power at its output will be (remembering that kT is 204 dB below 1 watt per cycle per second of bandwidth) 156 dB below 1 watt per cycle per second, or 86 dB below 1 watt in a 10 Mc/s bandwidth. If the saturated output of the tube is only, say, 500 μW, which is 33 dB below 1 watt, such a tube working into a receiver with a 10 Mc/s bandwidth would be able to handle signals lying in the 53 dB range between those which it can just detect and those sufficiently large to saturate it. This is known as the dynamic range of the tube.

The requirement for large dynamic range often accompanies the requirement for low noise, and the increased power required is inevitably accompanied by an increase of the noise figure above the minimum possible. This type of demand is generally met by ranges of broad-band p.p.m focused low-noise tubes, with saturated powers in the region of a few milliwatts. Noise figures are then in the range from about 8 dB at 2,000 Mc/s to 11 dB at 9,000 Mc/s.

For terrestrial systems, where the background noise coming into the aerial is roughly that corresponding to a source at about 290°K, there is little to be gained by reducing the receiver noise figure below about 3 dB, but for the reception of signals coming from outside the earth, such as in radio-astronomy, rocket telemetry or satellite-communication systems, the aerial may be looking through a relatively thin layer of atmosphere at a background with an effective temperature only a few degrees above absolute zero over the frequency range between 1 and 10 Gc/s. For this sort of reception a greatly reduced noise performance is clearly desired. This is usually expressed by referring to the excess noise temperature rather than to a noise figure, as mentioned earlier in the chapter.

For this sort of application, the low-noise travelling wave tube is, at least at present, clearly quite inadequate. Even the best ever attainable noise figure of 1 dB represents an effective noise temperature of 76°K, which relative to an effective aerial input temperature of, say, 5°K, represents a loss of 12 dB.

The basic trouble with any beam-type amplifier is that energy can only be obtained from the beam by interaction essentially with the slow space-charge wave, and once the beam has been set up the noise on this wave is irreducible. It will be recalled that the slow space-charge wave is increased in amplitude by taking power away from the beam, and so the only way to reduce the noise on this wave to zero would be to feed on to the beam noise which was exactly correlated with and in opposite phase to the noise already there. This is impossible, since the noise is, in fact, unique to that particular beam.

13.11 Parametric Amplification

With other waves, however, such as the fast space-charge wave, the position is different. Here a large wave amplitude represents extra energy in the beam, which may be removed by suitable couplers. This means that the noise energy of such a wave may be stripped from the beam before it enters an interaction region. As has been seen, interaction with this wave cannot be used to convert the d.c. beam energy into r.f. energy. What can be done, however, is to produce a large initial energy in the wave by coupling in r.f. power from an external generator, and then to use this energy to provide amplification for a signal. This process is called *parametric amplification*, and is of wide application in the field of low-noise amplifiers, although most practical examples are, in fact, solid-state devices.

The simplest type of parametric amplification, familiar to everyone, is that of a swinging pendulum excited by periodic pulling on its suspension. The oscillation of such a pendulum can be built up by decreasing its length a little when the bob is passing through the centre position, and increasing its length at the end of the swing. At the centre position the tension in the string is greater than the weight of the bob, because of the added centrifugal force. At the end of the swing the tension is less than the weight, because of the angular displacement. More work is therefore done in shortening the string than is regained when it is lengthened, and this excess work passes into the motion of the pendulum. It may be noticed that since the bob passes through the centre twice in each oscillation the applied drive is at twice the amplified frequency.

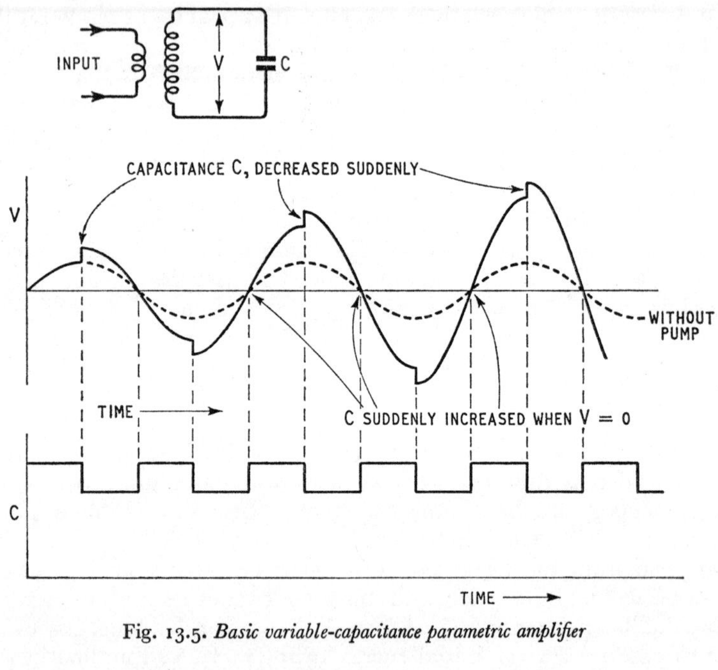

Fig. 13.5. *Basic variable-capacitance parametric amplifier*

The same principle is applied in the circuit of Fig. 13.5, in which it is supposed that the value of the capacitance is varied periodically. In the simplest case, in which it is varied at twice the signal frequency, if the capacitance is decreased at periods of maximum charge its voltage will rise and a larger current will flow during the next half-cycle. The capacitance is then restored during the period of minimum charge. Work has to be supplied to give this decrease of capacitance (e.g. in the simplest case of an air-spaced plate system the plates would have to be pulled apart against the electrostatic attraction) which provides the power source for the amplification.

For clarity, a pulsed change of capacitance has been considered, but a sinusoidal variation would also produce amplification, giving a more smoothly growing waveform than that shown in Fig. 13.5. The power source, in the above case at twice the signal frequency, is known as the pump. If it is not at twice the signal frequency, amplification can still take place, but the process is then accompanied by the production of a third frequency, called the 'idler' frequency, given by

$$f_{pump} = f_{signal} + f_{idler}$$

A circuit resonant at all three frequencies must then be used, as shown in Fig. 13.6.

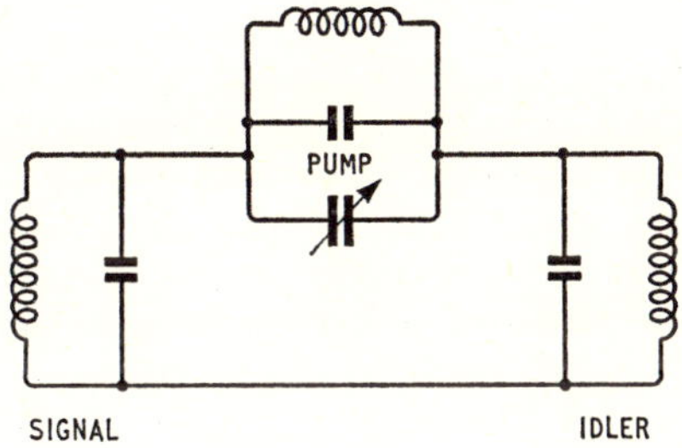

Fig. 13.6. *Non-degenerate parametric amplifier circuit*

It will be appreciated that in the two examples given, in which the pump signal is exactly twice that of the signal frequency, there is an essential phase relation between the signal and the pump power. This is called the degenerate case. In the more general non-degenerate case where the pump and signal frequencies are not in this relationship, there can be no such phase relation required, and the operation is therefore less critical.

13.12 The Transverse-Wave Parametric Amplifier

The only successful electron-beam amplifier which makes use of the parametric principle is the *transverse-wave amplifier, or Adler tube.* This uses the cyclotron motion of electrons in a drifting stream and has a general arrangement as shown in Fig. 13.7. The input coupler is essentially a pair of parallel plates, and the transverse r.f. electric

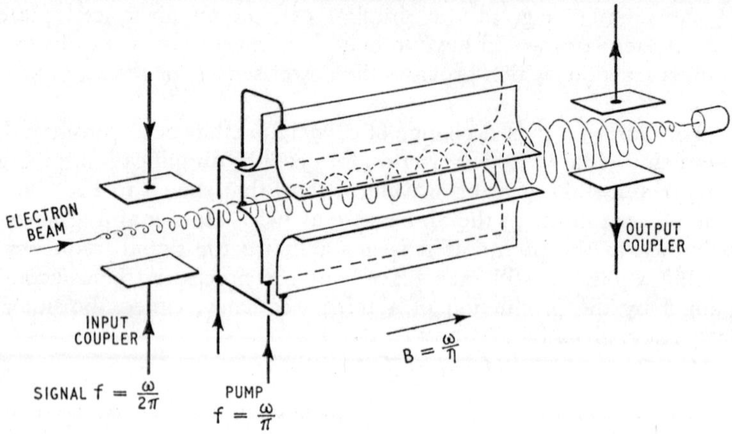

Fig. 13.7. *Electron beam transverse-wave amplifier*

field produced between these by the signal, in conjunction with the axial magnetic field, causes a circular movement of the electrons to build up as they pass through the coupler. The signal is then carried by this movement in the beam into the quadrupole pump system. This latter is fed from an external source at twice the signal frequency, but since the pump power is coupled on to the beam by a four-pole system the fields it produces are like those of a four-segment magnetron in its π-mode, and can be analysed as consisting of two fields rotating in opposite directions, of which one will therefore be in the same direction and at the same angular velocity as the signal. This therefore increases the amplitude of the circular motion of the beam, thus amplifying the signal, which is then coupled off the beam by a second transverse-field coupler at the output end. The electron gun of the device is a particular form of low-noise gun, with multiple grids to give the best velocity transformation from the cathode to the input coupler, but in this valve, because of the fact that the interaction is not with the slow-space-charge wave, the remaining noise can be removed from the beam before the signal is coupled on to it. This is done by the input coupler, the coupled-out noise being fed out to the input source, which must therefore be matched. In the simplest case, the noise would be radiated from the receiving aerial.

Amplifiers of this sort have been successfully made in the region from a few hundred to a few thousand Mc/s, and have produced noise figures as low as 0·8 dB, with gains as high as 30 dB. Their chief limitation is the bandwidth of the coupler, and it has not so far been possible to find couplers with a greater bandwidth than about 10%.

13.13 Parametric Amplification by Varactor Diodes

The majority of examples of parametric amplification have been those using variable capacitance diodes, or *varactors*, and masers, and although these might be regarded as outside the scope of a book such as this, a brief description is included, since by means of such devices very low noise temperatures of about 20°K have been obtained.

The variable-capacitance diode is composed of two layers of semiconducting material in contact with each other. One layer might be a material such as *n*-type silicon which has an excess of conducting electrons, and the other *p*-type silicon which has the opposite characteristic, often referred to as an excess of holes. When a voltage is applied across the interface between the two types of material a small current flows, which rapidly becomes diminished by the field produced inside the material by the displacement of charges across the boundary. It has therefore behaved partially as a capacitor. The magnitude of the capacitance depends on the instantaneous value of applied voltage, and follows this so closely that capacitance variation takes place even at microwave frequencies. In their application to parametric amplifiers, varactor diodes enable the capacitance variation to be made entirely by electrical means, by the application of the pump r.f. voltage. The way in which such a diode is used is basically to place it in a circuit which has resonances at the signal, pump and idler frequencies, so that the maximum fields at these frequencies will be produced across the diode for the minimum amounts of power. Such a system might be a resonant cavity, which is arranged so that three of its resonant modes are at the required frequencies. The amplifier is then used in conjunction with a ferrite circulator, as indicated in Fig. 13.8, which directs the incoming signal power from the aerial into the cavity, and the outgoing amplified power from the cavity, which appears at the same coupling, into the receiver. If necessary, the whole amplifier can be cooled to a very low temperature, and by this means noise temperatures as low as 25°K have been obtained.

13.14 Masers

The *maser* (microwave amplification by stimulated emission of radiation), which is the other principal solid-state parametric

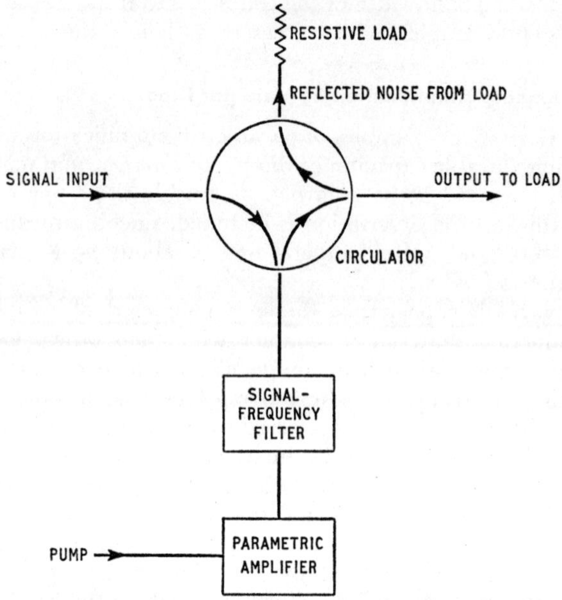

Fig. 13.8. *Parametric amplifier used with circulator*

amplifier, produces similar results by making use of energy levels in the molecule itself. In any molecule there are many possible orbits which the electrons may occupy, and in each of these the energy of the system is slightly different. The most likely state for a molecule to be in is that in which the energy is lowest, and this is called the 'ground state'. If it is excited to a higher energy level, e.g. by collision with another particle, or by absorbing some radiation falling upon it, it will, sooner or later, return or 'relax' to its ground state, and will emit some radiation of just the same energy as it does so. The most familiar example of this is probably that of a discharge lamp, such as the orange sodium discharge lamps used for street lighting. In these the molecules of sodium vapour are constantly being excited by collisions in the discharge set up in the lamp, and when reverting to their ground state give out small quantities of light, which because of the constant difference between the energy E_1 of the excited state and that, E_2, of the ground state are always at the same frequency, given by

$$hf = E_1 - E_2$$

Since all these molecules are behaving independently, the overall light output is not coherent, like that of a continuous oscillator, but random.

In a maser a material is selected in which levels can be found which are suitably spaced so that the frequencies involved in excitation and relaxation are in the microwave region. This requires energy levels which are quite close together, and these are often obtained by the splitting of single energy levels, which frequently occurs under the action of applied magnetic fields. This is also convenient, since the level difference is dependent on the applied field, so that a method of tuning is available. The material used is, for instance, ruby or one of the garnets, and this is mounted in a device in essentially the same way as a parametric diode.

As an example, we may consider the three-level maser, although masers involving two and four adjacent energy levels have also been made. In the three-level maser the material and magnetic field are chosen so that there are three adjacent energy levels which are separated by intervals so that the frequencies involved in the transitions are in the microwave region. If the energies of these states are denoted by E_1, E_2 and E_3, then

$$E_3 - E_1 = hf_1$$
$$E_3 - E_2 = hf_2$$
$$E_2 - E_1 = hf_3$$

If E_1 is the ground state having the lowest energy, this is the state that most of the molecules will initially be in, the population in a level with energy E_m is given by the well-known Boltzmann law

$$N_m = k'e^{-\frac{E_m}{kT}}$$

where k' is a constant.

When power at a pump frequency f_1 is applied, many molecules become excited to the state E_3, and an artificial population density is obtained. Then, if matters are suitably arranged, these molecules will decay to state E_2, giving out quanta of radiation at frequency f_2 as they do so. We now have a crystal with many molecules in the state E_2. When the crystal is subjected to the influence of fields at the frequency f_3, which it is desired to amplify, these molecules are disturbed and decay to their ground state E_1, giving out quanta at the signal frequency f_3 as they do so, and because of the nature of the stimulation of the decay the emitted radiation is in a constant phase relationship with that of the stimulating signal, so that coherent amplification is thus obtained.

As with the parametric amplifier, we see that the amplification has been accompanied by the production of a third frequency given by

$$f_1 = f_2 + f_3$$

but the emission at the unwanted frequency is not now coherent. As with the parametric diode, the whole system may be cooled to a very low temperature, so that random transitions, which in this system would constitute noise, occur only rarely, and very low noise temperatures of a few degrees may thereby be obtained.

Other types of low-noise solid-state amplifiers are used, and parametric diodes and masers are made and used in other ways than are described above, but their detailed treatment would be out of place here. It is hoped, however, that the above is sufficient to give some idea of the performances which solid-state devices can offer in the low-noise field. It will also be apparent that, as described, they are essentially very narrow-band devices. They are also, although this may be less obvious, very low-power devices, and generally have a lower power range and gain than can be obtained, for instance, from low-noise travelling wave tubes. It is therefore not uncommon to find in applications where very low noise figures are required, such as in the receiver of a satellite communication station, a maser amplifier, placed as near as possible to the aerial focus, and cooled by liquid helium, feeding into a low-noise travelling wave tube in the transmitter.

BIBLIOGRAPHY

VAN DER ZIEL, A., *Noise*, Chapman and Hall (1955).
SMULLIN, L. D., and HAUS, H. A., *Noise in Electron Devices*, Chapman and Hall (1959).

INDEX